Selected Titles in This Series

(Continued in the back of this publication)

Smooth Molecular Decompositions of Functions and Singular Integral Operators

MEMOIRS
of the
American Mathematical Society

Number 742

Smooth Molecular Decompositions of Functions and Singular Integral Operators

J. E. Gilbert
Y. S. Han
J. A. Hogan
J. D. Lakey
D. Weiland
G. Weiss

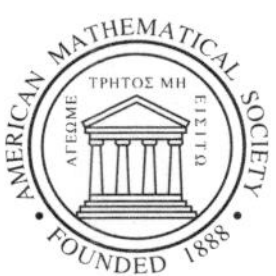

March 2002 • Volume 156 • Number 742 (third of 5 numbers) • ISSN 0065-9266

American Mathematical Society
Providence, Rhode Island

2000 *Mathematics Subject Classification.*
Primary 42B20, 42C15, 42C40.

Library of Congress Cataloging-in-Publication Data

Smooth molecular decompositions of functions and singular integral operators / J. E. Gilbert ... [et al.].
    p. cm. — (Memoirs of the American Mathematical Society, ISSN 0065-9266 ; no. 742)
"Volume 156, number 742 (third of 5 numbers)."
Includes bibliographical references.
ISBN 0-8218-2772-3
    1. Function spaces. 2. Integral operators. 3. Decomposition (Mathematics) I. Gilbert, John E. II. Series.
QA3.A57 no. 742
[QA323]
510s—dc21
[514$'$.322]
        2001056088

Memoirs of the American Mathematical Society

This journal is devoted entirely to research in pure and applied mathematics.

Subscription information. The 2002 subscription begins with volume 155 and consists of six mailings, each containing one or more numbers. Subscription prices for 2002 are \$524 list, \$419 institutional member. A late charge of 10% of the subscription price will be imposed on orders received from nonmembers after January 1 of the subscription year. Subscribers outside the United States and India must pay a postage surcharge of \$31; subscribers in India must pay a postage surcharge of \$43. Expedited delivery to destinations in North America \$35; elsewhere \$130. Each number may be ordered separately; *please specify number* when ordering an individual number. For prices and titles of recently released numbers, see the New Publications sections of the *Notices of the American Mathematical Society.*

Back number information. For back issues see the *AMS Catalog of Publications.*

Subscriptions and orders should be addressed to the American Mathematical Society, P. O. Box 845904, Boston, MA 02284-5904. *All orders must be accompanied by payment.* Other correspondence should be addressed to Box 6248, Providence, RI 02940-6248.

Memoirs of the American Mathematical Society is published bimonthly (each volume consisting usually of more than one number) by the American Mathematical Society at 201 Charles Street, Providence, RI 02904-2294. Periodicals postage paid at Providence, RI. Postmaster: Send address changes to Memoirs, American Mathematical Society, P. O. Box 6248, Providence, RI 02940-6248.

Contents

ABSTRACT. Under minimal assumptions on a function ψ we obtain wavelet-type frames of the form

$$\psi_{j,k}(x) \;=\; r^{\frac{1}{2}nj}\psi(r^j x - sk) \qquad j \in \mathbb{Z}, k \in \mathbb{Z}^n,$$

for some $r > 1$ and $s > 0$. This collection is shown to be a frame for a scale of Triebel-Lizorkin spaces (which includes Lebesgue, Sobolev and Hardy spaces) and the reproducing formula converges in norm as well as pointwise a.e. The construction follows from a characterization of those operators which are bounded on a space of smooth molecules. This characterization also allows us to decompose a broad range of singular integral operators in terms of smooth molecules.

CHAPTER 1

Main results

Introduction.[1]

The historical development of Calderón-Zymund theory, especially as it applies to end-point spaces and various geometrical settings, has proceeded along two main fronts. The first was concerned with understanding oscillation properties needed to establish basic boundedness results, the $T(1)$-theorem being perhaps the most fundamental example, while the P_t-Q_t methods of Coifman and Meyer replaced subtle oscillations restrictions with more effective Carleson measure conditions. The second front dealt with basic building blocks for function spaces and operators, including the representation of Calderón-Zygmund operators as well as their action on the building blocks themselves ([**MeYb, FJWa**]). For instance, the Calderón Reproducing formula provided a continuous decomposition of the identity operator; atoms, molecules and wavelets, on the other hand, were used to provide corresponding discrete decompositions, allowing questions of boundedness to be rephrased in terms of more transparent matrix boundedness. Underlying all these developments, either explicitly or implicitly, was the action of the affine group.

The introduction of wavelets did not lead, however, to a full understanding of the relationship between the distributional kernels of operators and their representations via various families of molecular building blocks. This occurred for several reasons. Being a basis is too strong a condition to impose on a family of functions - a more flexible, intrinsic restriction is needed. Secondly, the Lipschitz condition traditionally imposed on the kernel of a Calderón-Zygmund operator is not precise enough to guarantee that the operator preserves fully the fundamental properties of such families of building blocks - there is an inevitable loss of smoothness in general. In this memoir we shall address both issues, developing at the same time a theory of smooth decompositions of singular integral operators which unifies all the previous approaches. It employs techniques that are independent of L^2-theory and of Fourier transforms. The requisite flexibility and Lipschitz smoothness are achieved

- through the use of *frame decompositions* with basic building blocks drawn from a Banach space $\mathcal{M}_\delta$ of molecules whose decay, smoothness and cancellation depend on a parameter δ, $0 < \delta < \infty$, and
- by imposing *double* Lipschitz bounds on the kernel of Calderón-Zygmund operators formulated in terms of a so-called $\mathcal{H}_\delta$-*condition*.

At the expense possibly of some repetition the main results of the memoir will be described in this chapter within the context of known results. Full statements of definitions, results and detailed proofs then appear in the succeeding chapters.

---

[1]Received by the editor on Oct. 22, 1997. Revised manuscript received on Sep. 12, 2000

An axiomatic development of frame decompositions of the identity operator in the context of integrable, irreducible, unitary representations of a locally compact group has been developed in [**FG1**], [**FG2**] and [**Gro**]. By specializing to the affine group, one obtains frame decompositions of classical Banach spaces on $\mathbb{R}^n$ by identifying these spaces as co-orbit spaces associated with Banach lattices on the upper half-plane $\mathbb{R}_+^{(n+1)}$. Smoothness of both the analyzing and synthesizing functions in the decomposition, a key feature of this memoir, however, cannot be guaranteed with this axiomatic approach.

The theory of molecules and their applications to singular integrals developed in this memoir began with two separate efforts, one centered at the University of Texas at Austin, the other at Washington University. Initially, each had a different emphasis, but for the most part there was a considerable overlapping of ideas. Once the extent of the overlap was realized, it was decided that one joint publication would best represent the efforts of the two groups.

1. Frame decompositions.

A countable family of vectors $\{\varphi_\lambda\}_{\lambda \in \Lambda}$ in a separable Hilbert space $\mathcal{H}$ is said to be a *frame* when there exist positive constants C_1 and C_2 such that

$$(1.1) \qquad C_1 \|g\|^2 \leq \sum_{\lambda \in \Lambda} |\langle g, \varphi_\lambda \rangle|^2 \leq C_2 \|g\|^2$$

for all $g \in \mathcal{H}$. Because of Parseval's identity a frame is a natural generalization of the notion of orthonormal basis. The constants C_1 and C_2 in (1.1) are called the *frame bounds* and the condition itself is a *stability* condition because the coefficients $\langle g, \varphi_\lambda \rangle$ depend continuously on g. Any f in $\mathcal{H}$ can be reconstructed from the φ_λ using the *frame operator*

$$\mathcal{L}f = \sum_{\lambda \in \Lambda} \langle f, \varphi_\lambda \rangle \varphi_\lambda \, .$$

associated with the $\{\varphi_\lambda\}_{\lambda \in \Lambda}$ ([**Dau, HeW**]). Indeed, (1.1) ensures that $\mathcal{L}$ is bounded and invertible on $\mathcal{H}$ with operator norms

$$\|\mathcal{L}^{-1}\| \leq 1/C_1, \qquad \|\mathcal{L}\| \leq C_2$$

controlled by the frame bounds; in fact, $\mathcal{L}$ is bounded and invertible on $\mathcal{H}$ if and only if (1.1) holds. Thus the vectors $\rho_\lambda = \mathcal{L}^{-1}\varphi_\lambda$ again belong to $\mathcal{H}$ and form a frame $\{\rho_\lambda\}_{\lambda \in \Lambda}$, called the *dual frame* to $\{\varphi_\lambda\}_{\lambda \in \Lambda}$. The reconstruction of f in terms of the φ_λ is then given by

$$f = \sum_{\lambda \in \Lambda} \langle f, \rho_\lambda \rangle \varphi_\lambda = \sum_{\lambda \in \Lambda} \langle f, \varphi_\lambda \rangle \rho_\lambda \, .$$

In particular, the set $\{\varphi_\lambda\}_{\lambda \in \Lambda}$ is complete in $\mathcal{H}$ in the sense that finite linear combinations of the φ_λ form a dense subspace of $\mathcal{H}$. So, just as with orthonormal bases, frames give rise to stable reproducing formulas.

To extend the definition of a frame to a Banach space $\mathcal{B}$ the equivalent version of (1.1) is not used because the ℓ^2-sequence norm on the coefficients does not always have an obvious replacement for a general $\mathcal{B}$ whereas invertibility of an operator always makes sense. Nevertheless, if $\mathcal{B}$ is assumed to be, say, a co-orbit space associated with a Banach lattice on a locally compact group, then a sequence space definition is possible (see [**Gro**]). Let $\{e_\lambda\}_{\lambda \in \Lambda}$ be a family of vectors lying in the intersection $\mathcal{B} \cap \mathcal{B}^*$ of $\mathcal{B}$ and its dual space $\mathcal{B}^*$. We say that $\{e_\lambda\}_{\lambda \in \Lambda}$ is a *frame* for $\mathcal{B}$ if the *frame operator*

$$\mathcal{L} : f \to \sum_{\lambda \in \Lambda} \langle f, e_\lambda \rangle e_\lambda$$

is bounded and invertible on $\mathcal{B}$; as a result there exist positive constants C_1 and C_2 such that

$$C_1 \|f\|_\mathcal{B} \le \|\mathcal{L}f\|_\mathcal{B} \le C_2 \|f\|_\mathcal{B} \qquad (f \in \mathcal{B})$$

and so the operator norms of $\mathcal{L}$ and $\mathcal{L}^{-1}$ are still controlled exactly as for Hilbert spaces. Each f in $\mathcal{B}$ can then be reconstructed from the e_λ by the series

$$f = \sum_{\lambda \in \Lambda} \langle f, (\mathcal{L}^{-1})^* e_\lambda \rangle e_\lambda = \sum_{\lambda \in \Lambda} \langle f, e_\lambda \rangle \mathcal{L}^{-1} e_\lambda .$$

In particular, the family $\{e_\lambda\}_{\lambda \in \Lambda}$ will be complete in $\mathcal{B}$ once it is shown that the series expansion converges in the norm of $\mathcal{B}$.

The frames studied in this paper are generated by translation and dilation. For a function $\psi : \mathbb{R}^n \to \mathbb{C}$ and real numbers $r > 1, s > 0$ set

$$(1.2) \qquad \psi_{j,k}(x) = r^{\frac{1}{2}jn} \psi(r^j x + ks) = \psi_{j,k}^{(r,s)}(x), \qquad (j \in \mathbb{Z}, k \in \mathbb{Z}^n);$$

for notational convenience we shall usually omit the *mesh parameters* (r, s) and write simply $\psi_{j,k}$ except when particular values of (r, s) enter in an essential way. The family $\{\psi_{j,k}\}_{j,k}$ is thus a natural generalization of the well-known construction of orthonormal wavelet bases $\{\Psi_{j,k}\}_{j,k}$ for $L^2(\mathbb{R})$ as *dyadic* translates and dilates $2^{\frac{j}{2}} \Psi(2^j x - k)$ of a wavelet Ψ. As in the wavelet case we will consider both $L^2(\mathbb{R}^n)$ and more general function spaces. In the case of $L^2(\mathbb{R})$ Fourier transform techniques were used by Daubechies to prescribe conditions on ψ under which the family $\{\psi_{j,k}\}_{j,k}$ is a frame for $L^2(\mathbb{R})$ ([**Dau**]). These conditions require the Fourier transform of ψ to decay sufficiently fast at the origin and at infinity. The existence of frames for spaces other than $L^2(\mathbb{R})$ is a more delicate matter, however. Consider, for example, the so-called Mexican hat function

$$\varphi(x) = \frac{2}{\sqrt{3}} \pi^{-\frac{1}{4}} (1 - x^2) e^{-\frac{x^2}{2}}, \qquad \hat{\varphi}(\xi) = \frac{2}{\sqrt{3}} \pi^{-\frac{1}{4}} \xi^2 e^{-\frac{\xi^2}{2}}$$

and correspondingly on $\mathbb{R}^n$. Daubechies' criterion ensures that the family of dyadic translates and dilates of φ is a frame for $L^2(\mathbb{R})$, but it is not known if this family is complete in $L^p(\mathbb{R})$ for any p other than $p = 2$, as Y. Meyer points out ([**Mey**], chapter 4).

To study frames in classical Banach spaces of functions reliance on Fourier transform techniques has to be abandoned. Our approach will be to impose decay, smoothness and oscillation conditions on the function itself, rather than impose restrictions on its Fourier transform. The advantage of doing this is that we can then study frames within scales of function spaces of which $L^2(\mathbb{R}^n)$ is one member; it also leaves open the possibility of working on spaces of homogeneous type more general than Euclidean space. Some regularity is needed if a function ψ is to generate a frame since the existence merely of an upper frame bound C_2 is not obvious. The following definition captures the requisite decay and smoothness in a convenient way. It will play a crucial role.

(1.3) DEFINITION. Fix $\delta > 0$, δ not an integer. A function $f = f(x)$ on $\mathbb{R}^n$ is said to be an $\mathcal{M}_\delta$-*test function* when it has continuous derivatives up to order $[\delta]$ such that

$$|D^\alpha f(x)| \;\leq\; \frac{\text{const.}}{(1 + |x|)^{n+\delta+|\alpha|}}$$

holds for all $|\alpha| \leq [\delta]$, and when its highest order derivatives D^ν, $|\nu| = [\delta]$, satisfy the extra Lipschitz condition

$$|D^\nu f(x + y) - D^\nu f(x)| \;\leq\; \text{const.} \, \frac{|y|^{\delta - [\delta]}}{(1 + |x|)^{n+2\delta}}$$

for all $x, y \in \mathbb{R}^n$ with $|y| \leq \frac{1}{2}(1 + |x|)$.

The set of all $\mathcal{M}_\delta$-test functions becomes a Banach space, denoted by $\mathcal{M}_\delta(\mathbb{R}^n)$, taking as norm the smallest constant that can appear in the inequalities of definition (1.3). A Schwartz function, for instance, is an $\mathcal{M}_\delta$-test function for each $\delta > 0$; on the other hand, minimally smooth examples having slow decay can be constructed using splines. Notice that $\mathcal{M}_\delta$-test functions could also be defined when δ is a positive integer by requiring f to have continuous derivatives up to order $\delta - 1$ such that the highest order derivatives satisfy a Lipschitz condition of order 1. Smoothness and decay of all these functions could be prescribed separately, but in the interest of economy of notation this feature will not be adopted because no use will be made of it in this memoir.

By themselves, smoothness and decay of the function ψ are not enough to guarantee boundedness of the frame operator

$$S f = \sum_{j,k} \langle f, \psi_{j,k} \rangle \, \psi_{j,k}.$$

When ψ is a Gaussian, for example, it is easy to show that S is unbounded on $L^2(\mathbb{R})$ (see also (1.4)). Indeed, a necessary and sufficient condition for an $\mathcal{M}_\delta$-test function ψ to give rise to a bounded frame operator on L^p, $1 < p < \infty$, is that ψ have vanishing moment, i.e., $\int \psi(x) \, dx = 0$. To establish boundedness on spaces other than L^p, however, a more sophisticated notion of cancellation is needed: an $\mathcal{M}_\delta$-test function having vanishing moments

$$\int_{\mathbb{R}^n} f(x) x^\alpha \, dx \;=\; 0, \qquad (|\alpha| \leq [\delta]),$$

up to maximal order $[\delta]$ allowed by decay will be said to be an $\mathcal{M}_\delta$-*molecule*. The family $\mathcal{M}_\delta^{(0)}(\mathbb{R}^n)$ of all such $\mathcal{M}_\delta$-molecules is a closed subspace of $\mathcal{M}_\delta(\mathbb{R}^n)$. After translation and dilation

$$f \longrightarrow \frac{1}{d^{\frac{1}{2}n}} f\!\left(\frac{x - x_0}{d}\right)$$

an $\mathcal{M}_\delta$-molecule has *center* x_0 and *width* d.

Although the molecular condition guarantees that the frame operator S is bounded, some 'non-degeneracy' condition must be imposed on ψ to ensure S is invertible, and hence that ψ generate a frame. A necessary condition for this can be phrased in terms of the following inequality due to C. Chui and X. Shi ($[\mathbf{CS}]$): *if $\{\psi_{j,k}^{(r,s)}\}$ is a frame for $L^2(\mathbb{R}^n)$ then there exist constants A and B such that*

$$(1.4) \qquad 0 < A \; \leq \; \int_0^\infty |\hat{\psi}(t\xi)|^2 \, \frac{dt}{t} \; \leq \; B,$$

holds for almost all $\xi \in \mathbb{R}^n$. Notice that (1.4) will be satisfied when $\hat{\psi}$ is a continuous function such that

$$\operatorname{supp}\hat{\psi} \; \subseteq \; \{\xi : \tfrac{1}{2} \leq |\xi| \leq \tfrac{5}{2}\}, \qquad |\hat{\psi}(\xi)| > 0 \quad (1 \leq |\xi| \leq 2).$$

In this guise the Chui-Shi inequality is a requirement that ψ satisfy a Littlewood-Paley type inequality.

The function spaces for which we obtain decompositions can all be expressed in terms of the single scale of Triebel-Lizorkin spaces $\dot{F}_p^{\alpha,q}$. Here the index α measures smoothness, while the index p measures size; the index q is also a size index, but is of secondary importance (*cf.* chapter 2 for details). This well-known scale of spaces contains many classical function spaces:

$$L^2(\mathbb{R}^n) = \dot{F}_2^{0,2}, \quad L^p(\mathbb{R}^n) = \dot{F}_p^{0,2}, \quad H^1(\mathbb{R}^n) = \dot{F}_1^{0,2};$$

it includes Sobolev spaces also since $W_p^\alpha(\mathbb{R}^n) = \dot{F}_p^{\alpha,2}$.

We are now ready to state our main result concerning the frame decomposition of the Triebel-Lizorkin spaces $\dot{F}_p^{\alpha,q}$. It turns out that the necessary condition of Chui-Shi is also sufficient for small enough translates and dilates.

(1.5) THEOREM. *Let ψ be an $\mathcal{M}_\delta$-molecule satisfying (1.4), and fix γ, $\gamma < \delta$. Then there exist $r > 1$, $s > 0$, and a family of $\mathcal{M}_\gamma$-molecules $\{\rho_{j,k}\}$ such that*

$$f \; = \; \sum_{j,k} \langle\, f, \psi_{j,k}\,\rangle\, \rho_{j,k} \; = \; \sum_{j,k} \langle\, f, \rho_{j,k}\,\rangle\, \psi_{j,k}, \qquad (\psi_{j,k} = \psi_{j,k}^{(r,s)}),$$

for every f in $\dot{F}_p^{\alpha,q}$ provided $1 \leq p, q < \infty$ and $|\alpha| < \gamma$, the series converging in norm in $\dot{F}_p^{\alpha,q}$.

Although frame decompositions have been obtained for all of these spaces before, they have not been obtained with such minimal assumptions on the function ψ generating the frame. There is also a corresponding result for some indices p, q below 1 (*cf.* section 2.5). The example of the Mexican hat function φ illustrates (1.5) well. Since φ clearly satisfies (1.4), theorem (1.5) applies. Consequently, the family

$$\varphi_{j,k}^{(r,s)}(x) = r^{\frac{1}{2}jn}\, \varphi(r^j x + sk), \qquad (j \in \mathbb{Z}, \; k \in \mathbb{Z}^n)$$

is complete in $L^p(\mathbb{R}^n)$, $1 < p < \infty$, for sufficiently small $r > 1$ and $s > 0$, even though in the dyadic case, *i.e.*, when $r = 2$ and $s = 1$, it is not known if it is complete in any L^p-space other than $p = 2$.

We establish also a pointwise convergence result assuming only mild decay conditions on the dual frame elements:

$$(\ddagger) \qquad |\rho_{j,k}^{(r,s)}(x)| \; \leq \; \text{const.} \; \frac{r^{-\frac{1}{2}nj}}{(1 + |r^j x - sk|)^{n+\epsilon}}$$

holds uniformly in j, k for some $\epsilon > 0$. The inequality certainly holds for the $\rho_{j,k}$ in (1.5), for instance, since they are uniformly bounded in the $\mathcal{M}_\gamma$-norm. The following result thus establishes pointwise convergence for the frame decompositions in theorem (1.5) and more generally.

(1.6) THEOREM. *Suppose ψ is an $\mathcal{M}_\delta$-molecule that generates a frame for $L^2(\mathbb{R}^n)$. Then, if the dual frame elements $\{\rho_{j,k}\}$ satisfy ($\ddagger$), the reconstructions*

$$f = \sum_{j,k} \langle f, \psi_{j,k} \rangle \rho_{j,k} = \sum_{j,k} \langle f, \rho_{j,k} \rangle \psi_{j,k}$$

converge at every Lebesgue point of f.

In the very special case of (1.5) when ψ is a wavelet the corresponding decompositions are just orthonormal wavelet expansions which are known to converge in norm for this range of Triebel-Lizorkin spaces ([**Mey**]). Pointwise convergence results in the wavelet setting are known also ([**KKR**]).

Our proof of theorem (1.5) constructs frames using discretizations of the Calderón reproducing formula. Recall that a function ψ is said to be *admissible* if it is integrable and its Fourier transform is normalized in the sense that

$$\int_0^\infty |\hat{\psi}(t\xi)|^2 \, \frac{dt}{t} = 1, \qquad (\xi \neq 0)$$

([**Dau**]). Any such ψ necessarily has vanishing moment since finiteness of the integral forces $\hat{\psi}(0)$ to be zero. Calderón's reproducing formula then represents any $f \in L^2(\mathbb{R}^n)$ as the integral

$$(1.7) \qquad f = \int_0^\infty \psi_t * \tilde{\psi}_t * f \, \frac{dt}{t}$$

with respect to an admissible function ψ where $\tilde{\psi}(x) = \overline{\psi(-x)}$ and $\psi_t(x) = t^{-n}\psi(x/t)$ is the usual L^1-dilate of ψ. Formally, this can be seen by applying the Fourier transform to both sides of (1.7). Cotlar's lemma ensures that the integral in (1.7) converges in $L^2(\mathbb{R}^n)$, while convergence in $L^p(\mathbb{R}^n)$ and pointwise convergence of the integral were established by Saeki ([**Sae**]). It is well-known that appropriate discretizations of (1.7) lead to frame decompositions ([**FJW**]).

If instead of admissibility we assume only that ψ is an $\mathcal{M}_\delta$-molecule satisfying the Chui-Shi condition

$$0 < A \leq \int_0^\infty |\hat{\psi}(t\xi)|^2 \, \frac{dt}{t} \leq B, \qquad (\xi \in S^{n-1}),$$

in (1.4), then Calderón's formula does not apply directly. Nonetheless, it will allow us to recover a suitable reproducing formula. Define an operator T by

$$Tf = \int_0^\infty \psi_t * \tilde{\psi}_t * f \, \frac{dt}{t}.$$

Condition (1.4) ensures that T is bounded and invertible on $L^2(\mathbb{R}^n)$. Consequently, since T, and hence T^{-1}, commutes with translations and dilations we obtain the reproducing formula

$$(1.8) \qquad f = \int_0^\infty \tau_t * \tilde{\psi}_t * f \, \frac{dt}{t},$$

setting $\tau = T^{-1}\psi$. The problem with (1.8) is that it is not clear that τ has the same molecular properties as ψ. As noted in the Introduction, however, a Calderón-Zygmund operator does not in general preserve spaces of molecules but will do so if the Lipschitz condition imposed on its distributional kernel is strengthened as discussed in the next section. With this we can pass to the inverse of T using the following theorem proved in chapter 3.

(1.9) THEOREM. *Let T be a bounded, invertible operator on $L^2(\mathbb{R}^n)$ which commutes with translations and dilations. Then, if T is bounded on $\mathcal{M}_\delta(\mathbb{R}^n)$, its inverse T^{-1} will be bounded on $\mathcal{M}_\gamma(\mathbb{R}^n)$ for each $\gamma < \delta$.*

As with most theorems of this type, it is proved using Banach algebra techniques. In addition to the molecular results from the next section, the proof requires also a new characterization of Lipschitz spaces on the sphere (*cf.* (3.4.5)).

Returning to the proof of theorem (1.5), we can now discretize the modified Calderón reproducing formula (1.8): the limit

$$\lim_{(r,\,s)\to(1,\,0)} \left\{ s^n \ln r \left(\sum_{j,k} \langle f, \psi_{j,k}^{(r,s)} \rangle \tau_{j,k}^{(r,s)} \right) \right\} = \int_0^\infty \tau_t * \tilde{\psi}_t * f \, \frac{dt}{t} = f$$

exists at least formally. To make this precise, choose a sequence $\{(r_\ell,\, s_\ell)\}_\ell$ of mesh parameters such that

$$\lim_{\ell\to\infty} (r_\ell,\, s_\ell) = (1,\, 0)$$

and define $\mathcal{S}^{(\ell)}$ for an arbitrary pair of $\mathcal{M}_\delta$-molecules ϕ, ψ by $\mathcal{S}^{(\ell)}f = \mathcal{S}^{(r_\ell,\,s_\ell)}f$ where

$$(\dagger) \qquad\qquad \mathcal{S}^{(r,s)}f = (s^n \ln r) \sum_{j,k} \langle f, \psi_{j,k}^{(r,s)} \rangle \phi_{j,k}^{(r,s)}.$$

In chapters 3 and 4 using fully the action of the affine group on $\mathcal{M}_\delta(\mathbb{R}^n)$ and $L^p(\mathbb{R}^n)$ we shall prove the following results.

(1.10) THEOREM. *Let ϕ, ψ be $\mathcal{M}_\delta$-molecules and $\mathcal{S}^{(\ell)}$ the corresponding discrete operator ($\dagger$) determined by a sequence of mesh parameters $(r_\ell,\, s_\ell) \to (1,0)$. Then the maximal operator*

$$\mathcal{S}_* f(x) = \sup_\ell \left| \mathcal{S}^{(\ell)} f(x) \right|$$

is bounded on $L^p(\mathbb{R}^n)$. In particular,

$$\lim_{\ell\to\infty} \mathcal{S}^{(\ell)} f(x) = \int_0^\infty \phi_t * \tilde{\psi}_t * f(x) \, \frac{dt}{t}$$

exists both in norm in $L^p(\mathbb{R}^n)$ and pointwise a.e.

Applying this result to the particular molecules in (1.8) we see that

$$\mathcal{S}^{(r,s)} f(x) = (s^n \ln r) \sum_{j,k} \langle f, \psi_{j,k}^{(r,s)} \rangle \tau_{j,k}^{(r,s)}$$

defines a perturbation of the identity operator on $L^p(\mathbb{R}^n)$ for each choice of (r, s); in addition

$$\lim_{(r,s)\to(1,0)} (I - \mathcal{S}^{(r,s)}) = 0,$$

the limit being taken in the strong operator topology on $L^p(\mathbb{R}^n)$. Once this has been strengthened so that the limit exists in the *uniform operator* topology, we can choose (r, s) sufficiently small so that $\|I - \mathcal{S}^{(r,s)}\|_{\mathcal{L}(L^p)} < 1$ which via the usual Neumann series

$$(\mathcal{S}^{(r,s)})^{-1} = I + (I - \mathcal{S}^{(r,s)}) + (I - \mathcal{S}^{(r,s)})^2 + \cdots$$

ensures $\mathcal{S}^{(r,s)}$ will be invertible as an operator on $L^p(\mathbb{R}^n)$. Consequently, for a choice of sufficiently small mesh parameter (r, s),

$$(1.11)(i) \qquad\qquad f = \sum_{j,k} \langle f, \psi_{j,k}^{(r,s)} \rangle \, \rho_{j,k}^{(r,s)}$$

for all f in $L^p(\mathbb{R}^n)$ where

$$(1.11)(ii) \qquad\qquad \rho_{j,k}^{(r,s)} = s^n \ln r \, (\mathcal{S}^{(r,s)})^{-1} \tau_{j,k}^{(r,s)}$$

after absorbing the "over-sampling" constants $s^n \ln r$ into the dual frame elements $\rho_{j,k}^{(r,s)}$.

    This, on the surface, is the frame decomposition we have sought. But there are two crucial omissions. Firstly, theorem (1.5) deals with all Triebel-Lizorkin spaces $\dot{F}_p^{\alpha,q}$ in which $|\alpha| < \gamma < \delta$, and, secondly, the $\rho_{j,k}$'s have not been shown to inherit any of the smoothness and decay of ψ. On the other hand, it is known that an operator will bounded on $\dot{F}_p^{\alpha,q}$ once it is bounded on the space of $\mathcal{M}_\gamma$-molecules, $\gamma > |\alpha|$. Both omissions can thus be rectified simultaneously by establishing the following result.

    (1.12) THEOREM. *Let ψ, τ be the molecules appearing in the Calderón reproducing formula (1.8) and let*

$$\mathcal{S}^{(r,s)} f = (s^n \ln r) \sum_{j,k} \langle f, \psi_{j,k}^{(r,s)} \rangle \, \tau_{j,k}^{(r,s)}$$

be the associated perturbation operator. Then $\mathcal{S}^{(r,s)}$ is bounded on $\mathcal{M}_\gamma^{(0)}(\mathbb{R}^n)$ for each $\gamma < \delta$; furthermore,

$$\lim_{(r,s)\to(1,0)} \|I - \mathcal{S}^{(r,s)}\| = 0$$

in the uniform operator norm on $\mathcal{M}_\gamma^{(0)}(\mathbb{R}^n)$.

    By choosing (r, s) sufficiently small, therefore, we ensure that $\mathcal{S}^{(r,s)}$ is invertible both on $\mathcal{M}_\gamma^{(0)}(\mathbb{R}^n)$ and on $\dot{F}_p^{\alpha,q}$ provided $|\alpha| < \gamma$. In particular, the dual frame elements $\rho_{j,k}^{(r,s)}$ will be $\mathcal{M}_\gamma$-molecules and each f in $\dot{F}_p^{\alpha,q}$ will be given by (1.11), completing the proof of theorem (1.5) once all the intermediary results have been established. Let us now turn to the proof of these results. For this it will be convenient to broaden the discussion and consider the general question of molecular decomposition of operators and especially its connection with the affine group of transformations of Euclidean space.

2. Molecular boundedness and operator decompositions.

In the previous section the problem was that of representing a particular operator, the identity operator, as a sum or integral of translates and dilates of functions drawn from $\mathcal{M}_\delta(\mathbb{R}^n)$. The solution involved not just the identity operator but rather a whole family of operators defined by such sums and integrals. Their boundedness on spaces of molecules or, more generally, on L^p-spaces and Triebel-Lizorkin spaces was the critical issue. In this section we study these questions within the context of Calderón-Zygmund operators.

For a fixed $\delta > 0$, δ not an integer, let T be a linear operator that is bounded from the space $\mathcal{M}_\delta(\mathbb{R}^n)$ of test functions to its dual space, a space of distributions. Since T will then be continuous from the Schwartz space $\mathcal{S}(\mathbb{R}^n)$ to its dual, the Kernel theorem ensures that T is given formally as an integral operator

$$Tf(x) \; = \; \int_{\mathbb{R}^n} K(x,\, y)\, f(y)\, dy, \qquad (f \in \mathcal{S}).$$

The smoothness of $K = K(x,y)$ is used to categorize T. Recall that T is said to be a Calderón-Zygmund operator of order δ, written $T \in CZO(\delta)$, if on $\Omega = \{(x,y) : x \neq y\}$ the kernel K has continuous partial derivatives up to order $[\delta]$ in each variable satisfying the inequalities

$$\text{(i)} \qquad |(D_x^\alpha D_y^\beta K)(x,y)| \; \leq \; \text{const.} \; \frac{1}{|x - y|^{n+|\alpha|+|\beta|}},$$

$$\text{(ii)} \qquad |D_x^\nu D_y^\beta \{K(x+h, y) - K(x, y)\}| \; \leq \; \text{const.} \; \frac{|h|^{\delta-[\delta]}}{|x - y|^{n+\delta+|\beta|}}\; ,$$

$$\text{(iii)} \qquad |D_x^\alpha D_y^\nu \{K(x, y+k) - K(x, y)\}| \; \leq \; \text{const.} \; \frac{|k|^{\delta-[\delta]}}{|x - y|^{n+\delta+|\alpha|}}$$

for all $|\alpha|, |\beta| \leq [\delta]$, $|\nu| = [\delta]$, and $|h|, |k| \leq \frac{1}{2}|x - y|$. If, for instance, ϕ and ψ are $\mathcal{M}_\delta$-molecules, then both the 'integral' and 'discrete sum' operators

$$\text{(*)} \qquad f \; \longrightarrow \; \int_0^\infty \phi_t * \tilde{\psi}_t * f \, \frac{dt}{t}, \qquad f \; \longrightarrow \; \sum_{j,k} \langle f, \psi_{j,k}^{(r,s)} \rangle \, \phi_{j,k}^{(r,s)}$$

are Calderón-Zygmund operators of order γ for every $\gamma < \delta$. The behavior of $CZO(\delta)$-operators on molecules is well-known.

(2.1) THEOREM ([**Mey**]). *Let T be a Calderón-Zygmund operator of order ε, $0 < \varepsilon < 1$, such that*

    (i) T has the weak-boundedness property, and

    (ii) $T(1) = T^*(1) = 0$ modulo constants.

Then T is bounded from $\mathcal{M}_\gamma^{(0)}(\mathbb{R}^n)$ into $\mathcal{M}_{\gamma'}^{(0)}(\mathbb{R}^n)$ for each $\gamma' < \gamma < \varepsilon$.

For our purposes, theorem (2.1) needs to be strengthened before it can be applied to the Neumann series expansions discussed in the previous section. Indeed, suppose $T = I - S$ satisfies the hypotheses of (2.1). Then each iterate of $I - S$ decreases smoothness, and so all will be lost in the limit. Thus nothing could be said about the smoothness of $S^{-1}\tau$ no matter how smooth τ might be. Secondly we shall need the result for all $\delta > 0$, not just for $\delta < 1$. The requisite strengthening of (2.1) will come by imposing extra cancellation and smoothness conditions on the kernel $K = K(x, y)$. In chapter 2 the following result is established.

(2.2) THEOREM. *Let T be a Calderón-Zygmund operator of order δ, $\delta > 0$, such that*

> (i) *T has the weak-boundedness property, and*

> (ii) *$T(x^\alpha) = T(y^\alpha) = 0$ modulo polynomials of degree α for all $\alpha \le [\delta]$.*

Then T is bounded on $\mathcal{M}_\gamma^{(0)}(\mathbb{R}^n)$ for each $\gamma < \delta$ whenever $K = K(x, y)$ satisfies the double Lipschitz condition

$$\left| D_x^\mu D_y^\nu \left\{ K(x+h, y+k) - K(x, y+k) - K(x+h, y) + K(x, y) \right\} \right| \le \text{const.} \frac{(|h|\,|k|)^{\delta - [\delta]}}{|x - y|^{n+2\delta}}$$

for all $|\mu|, |\nu| = [\delta]$, and all $|h|, |k| \le \frac{1}{3}|x - y|$.

Weakening hypothesis (ii) produces a result of independent interest.

(2.3) THEOREM. *Let T be a Calderón-Zygmund operator of order δ, $\delta > 0$, such that*

> (i)   *T has the weak-boundedness property,*

> (ii)   *$T(y^\alpha) = 0$ modulo polynomials of degree α for all $\alpha \le [\delta]$, and*

> (iii)   *the kernel of T satisfies the double Lipschitz condition of Theorem (2.2).*

Then T is bounded from $\mathcal{M}_\gamma^{(0)}(\mathbb{R}^n)$ into $\mathcal{M}_\gamma(\mathbb{R}^n)$ for each $\gamma < \delta$.

The proof is virtually the same as for (2.2). In both results the norm of T as an operator can be controlled explicitly by the constants appearing in the definition of $CZO(\delta)$, the weak-boundedness property and the constant from the double Lipschitz condition on $K(x, y)$. When a Calderón-Zygmund operator satisfies the hypotheses of (2.3) we shall say that it *satisfies the $\mathcal{H}_\delta$-condition*. Two comments are in order. For $\delta < 1$ the hypotheses in (2.2) reduce to those in Meyer's result along with the requirement that the inequality

$$|K(x + h, y + k) - K(x, y + k) - K(x + h, y) + K(x, y)| \le \text{const.} \frac{|h|^\delta\,|k|^\delta}{|x - y|^{n+2\delta}}$$

holds for all $|h|, |k| \le \frac{1}{3}|x - y|$. Secondly, the explicit control on the operator norm will enable us to establish the norm bound $\|I - \mathcal{S}^{(r,s)}\|_{\mathcal{M}_\gamma^{(0)}} < 1$ for sufficiently small r and s, and hence invert $\mathcal{S}^{(r,s)}$ once we know that the kernel of $\mathcal{S}^{(r,s)}$ satisfies the double Lipschitz condition of (2.2). This can be checked directly given the specific structure of the kernel of $\mathcal{S}^{(r,s)}$, but the role of the affine group will be brought out much more clearly by exhibiting (essentially) the whole class of Calderón-Zygmund operators to which (2.2) and (2.3) apply since this class contains many of the singular operators of interest in analysis, not just the continuous and discrete sum versions in $(*)$.

3. Molecules and the affine group.

The emphasis on translation and dilation indicates the role of the affine group. Let H be the group of transformations $z = (v, t) : y \to y.z = ty + v$ of $\mathbb{R}^n$ generated by translation and dilation; the group structure on H is given by

$$w.z = (u, s).(v, t) = (tu + v, st), \qquad \begin{cases} u, v \in \mathbb{R}^n, \\ s, t > 0. \end{cases}$$

The inverse of $z = (v, t)$ is the pair $z^{-1} = (-v/t, 1/t)$. ¿From a practical point of view it is often very useful to think of H as the half-space

$$\mathbb{R}^{n+1}_+ = \{z : z = (v, t), v \in \mathbb{R}^n, t > 0\}.$$

Since translation and dilation commute with differentiation, each mapping $f(y) \to f(y.z)$ is bounded on $\mathcal{M}_\delta(\mathbb{R}^n)$. Consequently,

$$U(z) : f(y) \longrightarrow f(y.z) = t^{\frac{1}{2}n} f(ty + v) \qquad (z = (v, t))$$

defines a representation of H by bounded operators on $\mathcal{M}_\delta(\mathbb{R}^n)$; in particular, for each compact set $K \subset H \approx \mathbb{R}^{n+1}_+$ there is a constant C_K such that

$$(3.1) \qquad \|U(z)f\|_{\mathcal{M}_\delta} \leq C_K \|f\|_{\mathcal{M}_\delta} \qquad (z \in K).$$

Without extra smoothness on ψ, however, it is not true in general that $z \to U(z)$ is continuous on $\mathcal{M}_\delta$-test functions in the sense that

$$\lim_{z \to e} \|(I - U(z))\psi\|_{\mathcal{M}_\delta} = 0$$

as z converges to the identity $e = (0, 1)$ in H for the same reason that a bounded, continuous function on $\mathbb{R}^n$ need not be uniformly continuous. But the limit will be zero if it is taken in $\mathcal{M}_\gamma(\mathbb{R}^n)$ for any $\gamma < \delta$. This is basic to the proof of (1.12) (*cf.* sections 1, 2 in chapter 3)

The representation $z \to U(z)$ extends to any Lebesgue L^p-space or Hardy H^p-space containing $\mathcal{M}_\delta(\mathbb{R}^n)$ as a subspace. The same uniform boundedness on compact sets as in (3.1) will hold for all of these spaces, but, unlike the case of $\mathcal{M}_\delta(\mathbb{R}^n)$, the representation will be continuous on all Lebesgue and Hardy spaces except for L^∞. The *adjoint* $U^*(z)$ of $U(z)$ as an operator on $L^2(\mathbb{R}^n)$ is given by

$$U^*(z) : f(x) \longrightarrow \frac{1}{t^{\frac{1}{2}n}} f\left(\frac{x - v}{t}\right).$$

This is the key to the important link between the group H and Euclidean harmonic analysis. For instance, if ψ is a standard $\mathcal{M}_\delta$-molecule, meaning that it is centered at the origin and has unit width, then $U^*(z)\psi$, $z = (v, t)$, will be a smooth molecule of width t centered at v as this is traditionally defined. Conversely, if φ is smooth molecule of width t centered at v, then $U(z)\varphi$ will be a standard molecule. On the other hand, the mapping

$$(3.2) \qquad \mathcal{W}_\psi : f \longrightarrow \langle f, U^*(z)\psi \rangle$$

defined in terms of the inner product

$$\langle f, g \rangle \;=\; \int_{\mathbb{R}^n} f(x)\overline{g(x)}\,dx$$

on $L^2(\mathbb{R}^n)$ is just the *continuous wavelet transform* of f with respect to ψ; in the language of representation theory, (3.2) is a *matrix coefficient* of U. With this notation the Calderón reproducing formula can be rewritten as

$$f \;=\; \int_0^\infty \int_{\mathbb{R}^n} \langle f, U^*(z)\psi \rangle\, U^*(z)\psi\, \frac{dv\,dt}{t^{n+1}} \;=\; \int_H \langle f, U^*(z)\psi \rangle\, U^*(z)\psi\, dz$$

where dz is right Haar measure on the group H. To accommodate 'coefficients' such as those arising in paraproducts, let $a, b : H \to \mathcal{M}_\delta(\mathbb{R}^n)$ be bounded measurable functions and define $\mathcal{T}_{ab}$ formally by

$$\mathcal{T}_{ab} f \;=\; \int_H \langle f, U^*(z)a(z) \rangle\, U^*(z)b(z)\, dz.$$

If at least one of a, b has range in $\mathcal{M}_\delta^{(0)}(\mathbb{R}^n)$, then $\mathcal{T}_{ab}$ will be bounded from $\mathcal{M}_\delta(\mathbb{R}^n)$ to its dual space, but boundedness can be also be established in the absence of vanishing moments by imposing 'tent space' conditions on a, b; in any case, each $\mathcal{T}_{ab}$ will be a $CZO(\gamma)$-operator for all $\gamma < \delta$. Since the use of such factorizations of Calderón-Zygmund operators can be traced at least to Cotlar's thesis, we shall say that $\mathcal{T}_{ab}$ is of *Cotlar type* when $\mathcal{T}_{ab} : \mathcal{M}_\delta(\mathbb{R}^n) \to \mathcal{M}_\delta^*(\mathbb{R}^n)$ is bounded. The Calderón reproducing formula is simply the case where $a(z; x) = b(z; x) = \psi(x)$ for all z in H. Other examples are discussed in chapter 2. For instance, discretizations of the Calderón reproducing formula and paraproducts are of Cotlar type also.

Discrete sum operators arise from the choice of a *lattice* in H having a prescribed *mesh size*, hence the term *mesh parameter* used in section 1. For each fixed (r, s), $r > 1, s > 0$, choose an affine lattice

$$\lambda_{j,k} \;=\; (sk/r^j,\, 1/r^j), \qquad (j \in \mathbb{Z},\, k \in \mathbb{Z}^n),$$

in H and let Δ be the *fundamental tile*

$$\Delta \;=\; \{((y_1, \ldots, y_n), t) \in H : 0 \le y_m < s,\, 1 \le t < r\}$$

in H. Then the union $\bigcup_{j,k} \Delta.\lambda_{j,k}$ of right translates of Δ provides an *affine tiling* of H by tiles all having the same measure, *i.e.*, $|\Delta.\lambda_{j,k}| = |\Delta|$ independently of j and k because of the use of right invariant Haar measure. The importance of this affine lattice becomes evident when we observe that

$$U^*(\lambda_{j,k})\psi(x) \;=\; r^{\frac{1}{2}jn}\psi(r^j x - sk).$$

Fix $\mathcal{M}_\delta$-molecules ϕ, ψ and an ℓ^∞-sequence $\{\beta_{j,k}\}_{j,k}$, and define a, b by

$$a(z) \;=\; \sum_{j,k} \chi_\Delta(z\lambda_{j,k}^{-1}) U(z\lambda_{j,k}^{-1})\psi, \quad b(z) \;=\; \sum_{j,k} \beta_{j,k}\, \chi_\Delta(z\lambda_{j,k}^{-1}) U(z\lambda_{j,k}^{-1})\phi$$

where χ_Δ is the characteristic function of Δ. Clearly

$$\mathcal{T}_{ab} f \;=\; |\Delta| \sum_{j,k} \beta_{j,k} \langle f, \psi_{j,k} \rangle \phi_{j,k} \;=\; (s^n \ln r) \sum_{j,k} \beta_{j,k} \langle f, \psi_{j,k} \rangle \phi_{j,k}.$$

This reduces to a frame operator on specializing ψ, ϕ and $\{\beta_{j,k}\}_{j,k}$. More generally, let $\phi_{j,k}$ be a norm-bounded family of $\mathcal{M}_\delta$-molecules (or simply test functions), and set

$$b(z) \;=\; \sum_{j,k} \beta_{j,k}\, \chi_\Delta(z\lambda_{m,k}^{-1})U(z\lambda_{m,k}^{-1})\phi_{j,k}.$$

Then $\mathcal{T}_{ab}$ is of Cotlar type, and it reduces to the identity operator in special cases. The class of Cotlar type operators thus includes all the operators discussed earlier. What is more, the kernel of $\mathcal{T}_{ab}$ satisfies the double Lipschitz condition and the class of Cotlar type operators essentially characterizes those operators satisfying the hypotheses of theorems (2.2) and (2.3). More precisely, we prove

(3.3) THEOREM. *Let $a, b : H \longrightarrow \mathcal{M}_\delta^{(0)}(\mathbb{R}^n)$ be bounded measurable functions. Then the Cotlar type operator $\mathcal{T}_{ab}$ satisfies the hypotheses of theorem (2.2). Conversely, if T satisfies the hypotheses of (2.2), then for each $\gamma < \delta$ there exist bounded measurable functions $a, b : H \longrightarrow \mathcal{M}_\gamma^{(0)}(\mathbb{R}^n)$ such that $T = \mathcal{T}_{ab}$.*

(3.4) THEOREM. *Let $a : H \longrightarrow \mathcal{M}_\delta^{(0)}(\mathbb{R}^n)$, $b : H \longrightarrow \mathcal{M}_\delta(\mathbb{R}^n)$ be bounded measurable functions. Then the Cotlar type operator $\mathcal{T}_{ab}$ satisfies the hypotheses of theorem (2.3). Conversely, if T satisfies the hypotheses of (2.3), then for each $\gamma < \delta$ there exist bounded measurable functions $a : H \longrightarrow \mathcal{M}_\gamma^{(0)}(\mathbb{R}^n)$, $b : H \longrightarrow \mathcal{M}_\gamma(\mathbb{R}^n)$ such that $T = \mathcal{T}_{ab}$.*

As the final chapter in this development of Calderón-Zygmund theory regarded, on the one hand, in terms of kernels and cancellation conditions, and, on the other, in terms of superpositions of basic building blocks, we consider equi-convergence questions relating the two. Results in chapter 4 concern equi-convergence between operators obtained by truncating the kernels of Cotlar type operators and the ones obtained by truncating the space-scale parameters. These lead to pointwise convergence results of which (1.10) is just one example. Thus these results go well beyond known pointwise convergence results ([**KKR**]).

CHAPTER 2

Molecular decomposition of operators

By its construction the kernel of the frame operator has a very precise description in terms of $\mathcal{M}_\delta$-molecules; as we shall see, it also behaves well under dilation and translation even though it does not commute with such transformations. These properties will become important when the invertibility of the frame operator has to be determined. But the most convenient way of capturing these properties is by introducing a class of smooth Calderón-Zygmund operators which preserve the smoothness of a molecule as well as its center and width. In general, a Calderón-Zygmund operator decreases smoothness. When the kernel of the operator satisfies a Lipschitz condition jointly in both variables on the highest order derivatives, however, Y.S. Han showed that the operator preserves the smoothness of any molecule whose smoothness is less than that of the kernel. It is here that the definition of molecule adopted in the previous chapter plays a crucial role. By inverting the frame operator on this Banach space we obtain frame decompositions for a wide variety of classical function spaces.

The result is the realization of a large family of affine frame operators as CZO's having a generic form

$$\mathcal{T}_{ab} : f \longrightarrow \int_H \langle f, U^*(z)a(z)\rangle U^*(z)b(z)\, dz$$

$$= \int_0^\infty \left(\int_{\mathbb{R}^n} \langle f, U^*(z)a(z)\rangle U^*(z)b(z)\, dv \right) \frac{dt}{t^{n+1}} \qquad (z=(v,t))$$

where a,b are functions which are defined on the subgroup H of the Affine group of $\mathbb{R}^n$ generated by translations and dilations and which take values in a space on which $U(z)$ acts such as $\mathcal{M}_\delta$-test functions or, when cancellation is needed, a space of $\mathcal{M}_\delta$-molecules. Such a formalism includes the well-known P_t-Q_t methods. To make sense of the integral it has to be regularized. Traditionally this has been done through truncation in the spatial variable, though there are many other possibilities. This point of view leads to Cotlar-type operator theory in a variety of settings; vector-valued operators and operators on non-smooth domains are examples. Moreover this "pre-factorized" formulation lends itself to situations of interest, especially in representation theory, even when the operators are not necessarily CZO's.

1. Smooth Calderón-Zygmund operators

For fixed $\delta > 0$, let T be a linear operator that is bounded from the space of $\mathcal{M}_\delta$-test functions to its dual space in the sense that the inequality

$$(1.1) \qquad |\langle T\phi, \psi\rangle| \leq \mathrm{const.}\, \|\phi\|_{\mathcal{M}_\delta} \|\psi\|_{\mathcal{M}_\delta}$$

holds uniformly for all $\mathcal{M}_\delta$-test functions. Let $K(x,y)$ be its distribution kernel.

(1.2) DEFINITION. *An operator $T : \mathcal{M}_\delta(\mathbb{R}^n) \to \mathcal{M}_\delta^*(\mathbb{R}^n)$ satisfying (1.1) is said to be a Calderón-Zygmund operator of order δ if its kernel has continuous partial derivatives up to order $[\delta]$ in each variable on $\Omega = \{(x,y) : x \neq y\}$ satisfying the inequalities*

(i)
$$|D_x^\alpha D_y^\beta K(x,y)| \;\leq\; \text{const.} \; |x-y|^{-n-|\alpha|-|\beta|} \;,$$

(ii)
$$|D_x^\alpha D_y^\nu K(x,y) - D_y^\alpha D_x^\nu K(x',y)| \;\leq\; \text{const.} \; \frac{|x-x'|^{\delta-[\delta]}}{|x-y|^{n+\delta+|\alpha|}} \;,$$

(iii)
$$|D_x^\nu D_y^\beta K(x,y) - D_y^\nu D_x^\beta K(x,y')| \;\leq\; \text{const.} \; \frac{|y-y'|^{\delta-[\delta]}}{|x-y|^{n+\delta+|\beta|}}$$

for all $|\alpha|, |\beta| \leq [\delta]$ and $|\nu| = [\delta]$ whenever $x \neq y$ and $|x-x'|, |y-y'| \leq \frac{1}{2}|x-y|$.

We remark that if T is a Calderón-Zygmund operator of order δ and T is given by a principal value convolution with a kernel K, then K must have continuous partial derivatives of order $2[\delta]$ away from the origin. See section 4 of chapter 3 for a more detailed discussion.

We shall write $T \in CZO(\delta)$ for an operator satisfying the conditions in (1.2). Because of the symmetry in (1.1) and (1.2), the adjoint T^* of T belongs to $CZO(\delta)$ once T belongs, and the kernel of T^* is given by $\overline{K(y,x)}$ off the diagonal. On the other hand, as (1.2) is stronger than the definition adopted by Torres, all of his results apply ([**Tor**]). In particular, let $\mathcal{O}_\ell(\mathbb{R}^n)$ be the space of C^∞-functions having polynomial growth of degree at most ℓ, $\ell \leq [\delta]$. Then T extends to a bounded operator from $\mathcal{O}_\ell(\mathbb{R}^n)$ to the dual space of the space of $\mathcal{M}_\delta$-molecules so that (1.1) holds uniformly for all ϕ in $\mathcal{O}_\ell(\mathbb{R}^n)$ and all $\mathcal{M}_\delta$-molecules ψ; as a distribution $T(\phi)$ is then defined up to a polynomial. In this way $T(P)$ is well-defined for any polynomial $P = P(x)$ of degree at most $[\delta]$. We shall write $T(y^\ell) = 0$ when $T(P) = 0$ for all polynomials P of degree at most ℓ; similarly, $T^*(x^\ell) = 0$ will mean that $T^*(P) = 0$ for all polynomials P of degree at most ℓ.

(1.3) DEFINITION. *An operator T in $CZO(\delta)$ is said to satisfy the unilateral vanishing moments condition when $T(y^{[\delta]}) = 0$ or $T^*(x^{[\delta]}) = 0$, but not both, hold; by contrast, it will be said to satisfy the bilateral vanishing moments condition when $T(y^{[\delta]}) = 0$ and $T^*(x^{[\delta]}) = 0$ both hold.*

The dilation group H can be tied to any T in $CZO(\delta)$ through the action of $U(z)$ on $\mathcal{M}_\delta(\mathbb{R}^n)$ and $U^*(z)$ on the dual space $\mathcal{M}_\delta^*(\mathbb{R}^n)$. Since these last two operators are bounded for each z in H, the operator $U^*(z)TU(z)$ will be bounded from $\mathcal{M}_\delta(\mathbb{R}^n)$ to $\mathcal{M}_\delta^*(\mathbb{R}^n)$, and a simple calculation shows that the kernel of $U^*(z)TU(z)$ is given by

$$U^*(z) \otimes U^*(z)K(x,y) = t^{-n}K(x.z^{-1}, y.z^{-1}) \qquad (z = (v,t)).$$

Note that the kernel of $U^*(z)TU(z)$ will satisfy the same three conditions in (1.2) as the kernel of T does since these conditions are invariant under translation and dilation; furthermore, this invariance ensures that these conditions will be satisfied uniformly in $z \in H$. In summary, therefore,

(1.4) THEOREM. *If T is a Calderón-Zygmund operator of order δ, then so also is $U^*(z)TU(z)$; in addition, the estimates on the kernel of $U^*(z)TU(z)$ all hold uniformly in z.*

As the norm of the $U(z)$ as operators on $\mathcal{M}_\delta(\mathbb{R}^n)$ is not uniform in z, however, the norm of the mapping $U^*(z)TU(z) : \mathcal{M}_\delta(\mathbb{R}^n) \to \mathcal{M}_\delta^*(\mathbb{R}^n)$ need not be uniformly bounded in z. To require uniformity is one way of introducing the Weak Boundedness Property.

(1.5) DEFINITION. *An operator T in $CZO(\delta)$ is said to have the Weak Boundedness Property (WBP) if the inequality*

$$|\langle U^*(z)TU(z)\,\phi,\,\psi\,\rangle| \;\leq\; \text{const.}\,\|\phi\|_{\mathcal{M}_\delta}\,\|\psi\|_{\mathcal{M}_\delta} \qquad (z \in H)$$

holds uniformly in z for all ϕ, ψ in $\mathcal{M}_\delta(\mathbb{R}^n)$.

A more restrictive boundedness condition was introduced by Stein ([**Ste2**]): an operator T in $CZO(\delta)$ is said to have the *Restricted Boundedness Property (RBP)* when the inequality

$$\left(\int_{\mathbb{R}^n} |U^*(z)TU(z)\,\phi(x)|^2\,dx\right)^{1/2} \;\leq\; \text{const.}\,\|\phi\|_{\mathcal{M}_\delta} \qquad (z \in H)$$

holds uniformly in z for all ϕ in $\mathcal{M}_\delta(\mathbb{R}^n)$. In other words, T has WBP if $U^*(z)TU(z)$ is uniformly bounded in norm as an operator from $\mathcal{M}_\delta(\mathbb{R}^n)$ into its dual space $\mathcal{M}_\delta^*(\mathbb{R}^n)$, whereas T has RBP if $U^*(z)TU(z)$ is uniformly bounded in norm as an operator from $\mathcal{M}_\delta(\mathbb{R}^n)$ into $L^2(\mathbb{R}^n)$. The distinction between the two properties is brought out by the $T(1)$-theorem. For an operator T with WBP to have a bounded extension to $L^2(\mathbb{R}^n)$, one has to show that both $T(1)$ and $T^*(1)$ are BMO functions. By contrast, the advantage of the RBP definition is that T extends to a bounded operator on $L^2(\mathbb{R}^n)$ as soon as both T and its adjoint T^* have RBP (without reference to BMO).

The most precise molecular decompositions will be obtained when the operator *preserves* the smoothness of molecules. In general CZO-operators do not preserve smoothness, however. Nonetheless, as Meyer has shown ([**Mey2**]), if T belongs to $CZO(\delta)$ and has bilateral vanishing moments, then T maps molecules to molecules with minimal loss of smoothness. Unfortunately, the loss in smoothness, no matter how small, prevents anything useful being said about the smoothness that a power series $\sum_m a_m T^m$ preserves. Thus in general the usual Neumann series $\sum_{m\geq 0}(I - T)^m$ representation for the inverse of T cannot be used to deduce smoothness results for the inverse of T. What is needed is that $T(\psi)$ should have the *same* smoothness as ψ; ideally, $T(\psi)$ should also have the same center and width as ψ. Then there will be corresponding results for power series in T, and hence for T^{-1}. Extra conditions need to be imposed on the kernel of T. The appropriate condition in the case $0 < \delta < 1$, a *joint* Lipschitz condition on the kernel of T, was recognized first by Han ([**Han**]). Since we have allowed δ to have any positive, non-integral value, however, Han's result has to be extended to molecules of arbitrary smoothness. This will be done in section 4. The appropriate condition is then is a joint Lipschitz condition on the highest order derivatives of the kernel of T: the inequality

$$|D_x^\alpha D_y^\beta K(x,y) - D_x^\alpha D_y^\beta K(x',y) - D_x^\alpha D_y^\beta K(x,y') + D_x^\alpha D_y^\beta K(x',y')|$$

$$(1.6) \qquad\qquad \leq \text{const.}\,\frac{(|x-x'||y-y'|)^{\delta-[\delta]}}{|x-y|^{n+2\delta}} \qquad (|\alpha|,|\beta| = [\delta])$$

holds whenever $|x-x'|, |y-y'| \leq \tfrac{1}{3}|x-y|$. Because of the importance of (1.6) to this paper it will be convenient to introduce a class of $CZO(\delta)$-operators by collecting together in one definition all the conditions required to establish the preservation of smoothness. It is this preservation of smoothness that will then allow us to 'characterize' such operators as having a Cotlar type decomposition.

(1.7) DEFINITION. *An operator T in $CZO(\delta)$ is said to satisfy condition $\mathcal{H}(\delta)$ when*
    (i) *T has the Weak Boundedness Property,*
    (ii) *$T(y^{[\delta]}) = 0$,*
    (iii) *the kernel of T satisfies the joint Lipschitz condition (1.6).*
The smallest constant satisfying all the estimates in (1.2), (1.5) and (1.6) will be denoted by $\|T\|_{\mathcal{H}}$.

The generalization of Han's result now becomes

(1.8) THEOREM. *Let T be a $CZO(\delta)$-operator satisfying condition $\mathcal{H}(\delta)$ and fix γ, $\gamma < \delta$. Then T maps any $\mathcal{M}_\gamma$-molecule into an $\mathcal{M}_\gamma$-test function preserving center and width; more precisely, the inequality*

$$\|U^*(z)TU(z)(\psi)\|_{\mathcal{M}_\gamma} \ \leq \ \mathrm{const.}\,\|T\|_{\mathcal{H}}\,\|\psi\|_{\mathcal{M}_\gamma} \qquad (z \in H)$$

holds uniformly in T and z for all $\mathcal{M}_\gamma$-molecules ψ.

A proof of (1.8) will be given in section 4 of this chapter. For many applications it is important to know also that T preserves the space of $\mathcal{M}_\gamma$-molecules, *i.e*, T maps $\mathcal{M}_\gamma$-molecules to $\mathcal{M}_\gamma$-molecules, not just to $\mathcal{M}_\gamma$-test functions. For this we have to arrange that

$$\int_{\mathbb{R}^n} T\psi(x)\, x^\alpha \, dx = 0 \qquad (|\alpha| \leq [\delta]).$$

Clearly what is needed is that T^* also have vanishing moments. Hence

(1.9) COROLLARY. *Suppose that T satisfies $T^*(x^{[\delta]}) = 0$ in addition to the conditions in theorem (1.8). Then T maps any $\mathcal{M}_\gamma$-molecule into an $\mathcal{M}_\gamma$-molecule preserving center and width for $\gamma < \delta$.*

There is a partial converse to theorem (1.8).

(1.10) THEOREM. *Suppose an operator T maps $\mathcal{M}_\delta$-molecules into $\mathcal{M}_\delta$-test functions for some $\delta > 0$. Then $T \in CZO(\gamma)$ and satisfies condition $\mathcal{H}(\gamma)$ for any $\gamma < \delta$.*

Thus we can see that for a Calderón-Zygmund operator T, boundedness on molecular spaces and condition $\mathcal{H}$ are almost equivalent. To see that these conditions are not always satisfied consider the following example.

EXAMPLE. Let $K(x) = \max(0, 1 - |x - 1|)$ and define $Tf = K * f$. Since $K \in L^1$ it follows that T is bounded on L^2, and hence, $T \in WBP$. Since T is a convolution (see proposition 2.2.17 of [**Tor**]) it satisfies $T1 = T^*1 = 0$ in BMO. Moreover, it is easy to show that $T \in CZO(1)$. However, for $\frac{1}{2} < \varepsilon \leq 1$, T does not satisfy condition $\mathcal{H}(\varepsilon)$. Indeed, $K \in \Lambda_1$, the Lipschitz space of order 1, but no better. Thus K cannot satisfy a double-Lipschitz condition higher than $\frac{1}{2}$ (see lemma (4.5) in chapter 3). Therefore, by theorem (1.10), T cannot map $\mathcal{M}_\delta$-molecules into $\mathcal{M}_\delta$-test functions for $\delta > \frac{1}{2}$.

The $\delta > \frac{1}{2}$ appearing in this example is the most that we can say. In fact, we have the following result that says that any CZO operator satisfies a double-Lipschitz condition of smaller order.

(1.11) PROPOSITION. *Suppose $T \in CZO(\varepsilon)$, $0 < \varepsilon \leq 1$ with kernel K. Then*

$$\left|[K(x,y) - K(x',y)] - [K(x,y') - K(x',y')]\right|$$

$$\leq \text{const.} \, |x - x'|^{\frac{\varepsilon}{2}} |y - y'|^{\frac{\varepsilon}{2}} |x - y|^{-n-\varepsilon}$$

for $|x - x'|, |y - y'| \leq \frac{1}{3}|x - y|$.

PROOF. Since $|x - x'|, |y - y'| \leq \frac{1}{3}|x - y|$ it follows that $|x - x'| \leq \frac{1}{2}|x - y'|$. Thus

$$\left|[K(x,y) - K(x',y)] - [K(x,y') - K(x',y')]\right|$$

$$\leq |K(x,y) - K(x',y)| + |K(x,y') - K(x',y')|$$

$$\leq \text{const.} \, |x - x'|^\varepsilon |x - y|^{-n-\varepsilon} + \text{const.} \, |x - x'|^\varepsilon |x - y'|^{-n-\varepsilon}$$

$$\leq \text{const.} \, |x - x'|^\varepsilon |x - y|^{-n-\varepsilon}$$

since $|x - y| \sim |x - y'|$. Similarly we have

$$\left|[K(x,y) - K(x',y)] - [K(x,y') - K(x',y')]\right|$$

$$\leq |K(x,y) - K(x,y')| + |K(x',y) - K(x',y')|$$

$$\leq \text{const.} \, |y - y'|^\varepsilon |x - y|^{-n-\varepsilon} + \text{const.} \, |y - y'|^\varepsilon |x' - y|^{-n-\varepsilon}$$

$$\leq \text{const.} \, |y - y'|^\varepsilon |x - y|^{-n-\varepsilon} \, .$$

Taking the geometric mean of these two inequalities gives the desired result. $\square$

2. Cotlar type operators

Cotlar studied operators based on the translation and dilation structure common both to singular integral operators and to standard operators in ergodic theory ([**Cot**]). These operators are constructed in the following way. Let $y \to \sigma(y)$, $y \in \mathbb{R}^n$, be a group of measure preserving transformations of a measure space (X, μ) and let $f(x) \to f(\sigma(y)x)$ be the corresponding representation of $\mathbb{R}^n$ by unitary operators on $L^2(X, \mu)$. Then

$$(2.1) \qquad \mathcal{C}_{\phi\psi} : f \longrightarrow \int_0^\infty \int_{\mathbb{R}^n} (\psi_t * \phi_t^*)(y) f(\sigma(y)x) \frac{dy \, dt}{t} \qquad (x \in X)$$

formally associates an operator on $L^2(X, \mu)$ to any pair of functions ϕ, ψ on $\mathbb{R}^n$. In the prototypical case $\sigma(y) : x \to x - y$ of translation on $\mathbb{R}^n$ it becomes

$$(2.2) \qquad \mathcal{C}_{\phi\psi} : f \longrightarrow \int_0^\infty \int_{\mathbb{R}^n} \left(\int_{\mathbb{R}^n} \overline{\phi_t(y - v)} f(y) \, dy \right) \psi_t(x - v) \frac{dv \, dt}{t} \, .$$

If both of ϕ and ψ belong to $\mathcal{M}_\delta(\mathbb{R}^n)$ and at least one has vanishing moment to provide 'cancellation', then Cotlar's 'lemma' ensures that the integral in (2.2) exists in the sense that

$$f \longrightarrow \lim_{\varepsilon \to 0} \int_\varepsilon^\infty \int_{\mathbb{R}^n} \left(\int_{\mathbb{R}^n} \overline{\phi_t(y - v)} f(y) \, dy \right) \psi_t(x - v) \frac{dv \, dt}{t}$$

converges in norm in $L^2(\mathbb{R}^n)$ and defines a bounded operator on $L^2(\mathbb{R}^n)$.

Formally $\mathcal{C}_{\phi\psi}$ commutes with translation and dilation, *i.e.*, $\mathcal{C}_{\phi\psi}$ is invariant under the action of H in the sense that

$$(2.3) \qquad U^*(z)\mathcal{C}_{\phi\psi}U(z)f = \mathcal{C}_{\phi\psi}f \qquad (z \in H)$$

holds for all f on which $U(z)$ acts. The $\mathcal{C}_{\phi\psi}$ serve as the basic model for the operator of 'Cotlar type' through this structural dependence on the dilation group H introduced in section 1.3. Many operators of interest currently, however, cannot be represented by (2.2) because they do not commute with translation and dilation — bilinear singular integrals and wavelet or frame expansions, for instance. Nonetheless, they still maintain a structural dependence on translation and dilation, suggesting that the ϕ and ψ in (2.2) should be replaced by functions which preclude (2.3) but retain the essential translation and dilation structure of (2.2).

(2.4) DEFINITION. *Let* $a, b : H \to \mathcal{M}_\delta(\mathbb{R}^n)$ *be measurable,* L^∞-*bounded functions on* H *and define* $\mathcal{T}_{ab}$ *formally by*

$$\mathcal{T}_{ab} : f \longrightarrow \int_H \langle f, U^*(z)a(z) \rangle \, U^*(z)b(z)dz$$

$$= \int_H b(z\,;x.z^{-1})\left(\frac{1}{t^n}\int_{\mathbb{R}^n} \overline{a(z\,;y.z^{-1})}f(y)\,dy\right)dz,$$

the integral being taken with respect to right Haar measure $dz = dvdt/t^{n+1}$ *on* H.

When a, b are independent of z, in other words when $a(z\,;y) = \phi(y)$ and $b(z\,;x) = \psi(x)$, the corresponding $\mathcal{T}_{ab}$ will be said to be *homogeneous*. Such homogeneity, or lack of it, determines the invariance of $\mathcal{T}_{ab}$ under translation and dilation. Let

$$(2.5)\ (i) \qquad a_w(z\,;y) = a(zw\,;y) \qquad b_w(z\,;y) = b(zw\,;y) \qquad (w \in H)$$

be the translates of a, b with respect to the right regular representation of H; then, right invariance of Haar measure in (2.4) ensures, at least formally, that

$$(2.5)\ (ii) \qquad U(w)\mathcal{T}_{ab}U^*(w) = \mathcal{T}_{a_w b_w}, \qquad (w \in H)$$

so $\mathcal{T}_{ab}$ commutes with translation and dilation if and only if it is homogeneous, *i.e.* reduces to a $\mathcal{C}_{\phi\psi}$ operator.

The first basic problem, however, is to make sense of $\mathcal{T}_{ab}$. Set

$$(2.6)(i) \qquad \mathcal{T}_{ab}f(x) = \int_0^\infty \left(\int_{\mathbb{R}^n} K_t(x,y)f(y)\,dy\right)\frac{dt}{t}$$

where

$$(2.6)(ii) \qquad K_t(x,y) = \frac{1}{t^{2n}}\int_{\mathbb{R}^n} b(z\,;x.z^{-1})\overline{a(z\,;y.z^{-1})}\,dv \qquad (z = (v,t))\ .$$

To estimate such integrals the first of the simple Basic Estimates is needed.

BASIC ESTIMATE A(i). *The inequality*

$$\int_{\mathbb{R}^n} \frac{t^\delta}{(t+|x-v|)^{n+\delta}} \frac{t^\delta}{(t+|y-v|)^{n+\delta}}\,dv \le \text{const.}\ \frac{t^\delta}{(t+|x-y|)^{n+\delta}}$$

holds uniformly in x, y and t for each fixed $\delta > 0$.

A proof of A(i) appears in the Appendix. The decay built into $\mathcal{M}_\delta$-functions ensures that the outer integral in (2.6)(i) converges on (ε, ∞) for each $\varepsilon > 0$: indeed, set

$$(2.7)(i) \qquad \|a\|_{\infty,\delta} = \operatorname*{ess\,sup}_{z \in H} \|a(z\,;.)\|_{\mathcal{M}_\delta}, \qquad \|b\|_{\infty,\delta} = \operatorname*{ess\,sup}_{z \in H} \|b(z\,;.)\|_{\mathcal{M}_\delta},$$

and

$$(2.7)(ii) \qquad C_{ab} = \|a\|_{\infty,\delta} \|b\|_{\infty,\delta}.$$

Then, in view of A(i),

$$|K_t(x,y)| \leq C_{ab} \int_{\mathbb{R}^n} \frac{t^\delta}{(t + |x - v|)^{n+\delta}} \frac{t^\delta}{(t + |y - v|)^{n+\delta}}\, dv$$

$$(2.8) \qquad\qquad \leq \text{const.}\ C_{ab} \frac{t^\delta}{(t + |x - y|)^{n+\delta}}\ .$$

By Young's inequality, therefore,

$$\left(\int_{\mathbb{R}^n} \left| \int_{\mathbb{R}^n} K_t(x,y) f(y)\, dy \right|^2 dx \right)^{1/2} \leq \text{const.}\ C_{ab} \|f\|_{\mathcal{M}_\delta} \min(1, t^{-n/2})$$

and so

$$f \longrightarrow \int_\varepsilon^\infty \left(\int_{\mathbb{R}^n} K_t(x,y) f(y)\, dy \right) \frac{dt}{t}$$

defines a linear operator which is bounded from $\mathcal{M}_\delta(\mathbb{R}^n)$ into $L^2(\mathbb{R}^n)$ for each $\varepsilon > 0$. Just as in the homogeneous case, the last integral will not converge as $\varepsilon \to 0$ without further restrictions on a, b, but in contrast to the homogeneous case, 'cancellation' can take a variety of forms when a, b depend on z as the examples given later in this section show. Convergence as $\varepsilon \to 0$ will thus be built into the definition of an operator of Cotlar type.

(2.9) DEFINITION. *The operator $\mathcal{T}_{ab}$ associated with L^∞-bounded functions $a, b : H \to \mathcal{M}_\delta(\mathbb{R}^n)$ is said to be of Cotlar type when the integral in (2.6)(i) is well-defined in the sense that*

$$\langle \mathcal{T}_{ab} f,\, g \rangle := \lim_{\varepsilon \to 0} \int_\varepsilon^\infty \left(\int_{\mathbb{R}^n} \int_{\mathbb{R}^n} K_t(x,y) f(y) \overline{g(x)}\, dy\, dx \right) \frac{dt}{t}$$

exists for all f, g in $\mathcal{M}_\delta(\mathbb{R}^n)$ and $f \to \mathcal{T}_{ab} f$ defines a bounded operator from $\mathcal{M}_\delta(\mathbb{R}^n)$ into its dual space $\mathcal{M}_\delta^*(\mathbb{R}^n)$.

The construction of $\mathcal{T}_{ab}$ allows an explicit description of the distribution kernel of a Cotlar operator to be made. Set

$$(2.10) \qquad K(x,y) = \int_0^\infty K_t(x,y) \frac{dt}{t} = \int_H U^*(z) b(z;x)\, \overline{U^*(z) a(z;y)}\, dz\ .$$

By (2.8), the kernel K_t is in $L^1(0, \infty)$ uniformly in x, y so long as $\min_{x,y} |x - y| \geq \varepsilon'$ for some $\varepsilon' > 0$. Thus

$$\mathcal{T}_{ab}(f, g) = \int_{\mathbb{R}^n} \int_{\mathbb{R}^n} \left(\lim_{\varepsilon \to 0} \int_\varepsilon^\infty K_t(x, y) \frac{dt}{t} \right) f(y) \overline{g(x)} \, dy dx$$

$$= \int_{\mathbb{R}^n} \int_{\mathbb{R}^n} K(x, y) f(y) \overline{g(x)} \, dy \, dx$$

whenever f, g have disjoint compact supports, proving the following result.

(2.11) THEOREM. *Let $\mathcal{T}_{ab}$ be an operator of Cotlar type associated with L^∞-bounded functions $a, b : H \to \mathcal{M}_\delta(\mathbb{R}^n)$. Then off the diagonal the distribution kernel of $\mathcal{T}_{ab}$ coincides with the kernel $K = K(x, y)$ defined by (2.9).*

In the next section we shall see that $K(x, y)$ satisfies all the conditions of (1.2) for each fixed γ, $0 < \gamma < \delta$. Consequently, an operator of Cotlar type formed from functions $a, b : H \to \mathcal{M}_\delta(\mathbb{R}^n)$ is automatically a $CZO(\gamma)$-operator for such γ. Since each $U(w)$ is bounded on $\mathcal{M}_\delta(\mathbb{R}^n)$, the invariance condition (2.5) (ii) ensures that $\mathcal{T}_{a_w b_w}$ is also of Cotlar type whenever $\mathcal{T}_{ab}$ is. Thus individual operators of Cotlar type will not in general be invariant under H, but the *family* of Cotlar type operators will be invariant under H.

As might be expected, the integral defining $\mathcal{T}_{ab}$ converges when either a or b has vanishing moment, *i.e.*, whenever $a(z; \cdot)$ or $b(z; \cdot)$ is an $\mathcal{M}_\delta$-molecule for each z rather than just an $\mathcal{M}_\delta$-test function. In fact, in these cases much more can be said.

(2.12) THEOREM. *Let $K_t(x, y)$ be the kernel (2.6) associated with L^∞-bounded functions $a : H \to \mathcal{M}_\delta^{(0)}(\mathbb{R}^n)$, $b : H \to \mathcal{M}_\delta(\mathbb{R}^n)$. Then the inequality*

$$\int_0^\infty \left(\int_{\mathbb{R}^n} \left| \int_{\mathbb{R}^n} K_t(x, y) f(y) \, dy \right|^2 dx \right)^{1/2} \frac{dt}{t} \leq \text{const.} \, C_{ab} \|f\|_{\mathcal{M}_\gamma}$$

holds uniformly in a, b for all $f \in \mathcal{M}_\gamma(\mathbb{R}^n)$ where $\gamma < \delta$. In particular, $\mathcal{T}_{ab}$ is of Cotlar type.

If b has vanishing moment, then $\mathcal{T}_{ba}$ is of Cotlar type; in particular, the adjoint of $\mathcal{T}_{ab}$ also will be of Cotlar type since $\mathcal{T}_{ab}^* = \mathcal{T}_{ba}$. To exploit a vanishing moment condition on a or b, an estimate on the remainder term in Taylor's theorem for $\mathcal{M}_\gamma$-test functions is needed. Denote by

$$P_{[\gamma]} f(x, v) = \sum_{|\alpha| \leq [\gamma]} \frac{1}{\alpha!} D^\alpha f(x) v^\alpha$$

the Taylor polynomial of f of degree $[\gamma]$ centered at x.

BASIC ESTIMATE B(i). *Fix $\gamma > 0$, γ not a positive integer. Then for each δ, $\delta > \gamma$, the inequality*

$$\int_{\mathbb{R}^n} |f(x + sv) - P_{[\gamma]} f(x, sv)| \frac{1}{(1 + |v|)^{n+\delta}} \, dv$$

$$\leq \text{const.} \, \|f\|_{\mathcal{M}_\gamma} \frac{s^\gamma}{(1 + |x|)^{n+\gamma}}$$

holds uniformly in s, $s > 0$, for all $\mathcal{M}_\gamma$-test functions f.

A proof of B(i) appears in the Appendix.

PROOF OF THEOREM (2.12). The improvement over the earlier Young's inequality argument comes from using B(i) rather than A(i) when t is small. Fix γ with $[\delta] < \gamma < \delta$; since any $\mathcal{M}_\delta$-molecule is then an $\mathcal{M}_\gamma$-molecule, estimate B(i) applies. The vanishing moment condition on a ensures that

$$\frac{1}{t^n} \int_{\mathbb{R}^n} \overline{a(z\,;\,y.z^{-1})} f(y)\,dy = \int_{\mathbb{R}^n} \overline{a(z\,;\,y)} f(ty + v)\,dy$$

$$= \int_{\mathbb{R}^n} \overline{a(z\,;\,y)} \left\{ f(ty + v) - P_{[\gamma]} f(v, ty) \right\}\,dy.$$

Consequently, in view of B(i),

$$\left| \frac{1}{t^n} \int_{\mathbb{R}^n} \overline{a(z\,;\,y.z^{-1})} f(y)\,dy \right| \leq \text{const.} \|a\|_{\infty,\delta} \|f\|_{\mathcal{M}_\gamma} \frac{t^\gamma}{(1 + |v|)^{n+\gamma}}.$$

By Young's inequality, therefore,

$$\left(\int_{\mathbb{R}^n} \left| \int_{\mathbb{R}^n} K_t(x, y) f(y)\,dy \right|^2 dx \right)^{1/2} \leq \text{const.}\, C_{ab} \|f\|_{\mathcal{M}_\gamma}\, t^\gamma.$$

This, together with the corresponding estimate for large t, shows that

$$\left(\int_{\mathbb{R}^n} \left| \int_{\mathbb{R}^n} K_t(x, y) f(y)\,dy \right|^2 dx \right)^{1/2} \leq \text{const.}\, C_{ab} \|f\|_{\mathcal{M}_\gamma} \min(t^\gamma, t^{-n/2})$$

holds whenever a has vanishing moment. Boundedness of $\mathcal{T}_{ab}$ from $\mathcal{M}_\gamma(\mathbb{R}^n)$ into $L^2(\mathbb{R}^n)$ now follows immediately. $\qquad\square$

The first set of examples has appeared already and has been studied using P_t-Q_t methods. Conventional notation will be used: ϕ will be an $\mathcal{M}_\delta$-test function while ψ will be an $\mathcal{M}_\delta$-molecule.

(2.13) EXAMPLES. (i) (Para-products):

$$f \longrightarrow \int_0^\infty \int_{\mathbb{R}^n} \psi_t(x - v)\overline{(\psi_t^* * h)(v)}(\phi_t^* * f)(v)\, \frac{dv\,dt}{t}.$$

Here

$$a(z\,;\,y) = (\psi_t^* * h)(v)\phi(y), \qquad b(z\,;\,x) = \psi(x)$$

with $h \in BMO(\mathbb{R}^n)$. It satisfies the unilateral vanishing moment condition. The function $\psi_t^* * h$ will be bounded because ψ is in H^1.

(ii) (g-functions):

$$f \longrightarrow \int_{\mathbb{R}^n} \int_0^\infty m(t)\psi_t(x - v)h(v)(\phi_t^* * f)(v)\, \frac{dv\,dt}{t}.$$

Here

$$a(z\,;\,y) = h(v)\phi(y), \qquad b(z\,;\,x) = m(t)\psi(x)$$

with $h \in L^\infty(\mathbb{R}^n)$ and $m \in L^\infty(0, \infty)$. Again it has the unilateral vanishing moment property. As a form

$$\langle \mathcal{T}_{ab} f, g \rangle = \langle B(f, g), h \rangle \,,$$

consequently, this $\mathcal{T}_{ab}$-operator becomes the bi-linear singular integral

$$B(f, g)(v) = \int_0^\infty m(t)(\phi_t^* * f)(v)\overline{(\psi_t^* * g)(v)}\, \frac{dt}{t}$$

introduced by Coifman and Meyer.

(iii) An example which fails even the unilateral vanishing moment condition is

$$T(f, h)(x) = \int_{\mathbb{R}^n} \int_0^\infty m(t)\phi_t(x - v)\overline{(\psi_t^* * h)(v)}(\phi_t^* * f)(v)\, \frac{dvdt}{t}$$

with $m \in L^\infty(0, \infty)$ and $h \in BMO(\mathbb{R}^n)$. Here

$$a(z\,;\,y) = (\psi_t^* * h)(v)\phi(y) \,, \qquad b(z\,;\,x) = m(t)\phi(x).$$

The requisite 'cancellation' has been put into the function $\psi_t^* * h$ on H. This example is intimately connected with the bilinear form

$$\mathcal{T}(f, h) = \int_0^\infty m(t)\overline{(\psi_t^* * h)(v)}(\phi_t^* * f)(v)\, \frac{dt}{t}$$

whose L^2-boundedness was established by Coifman and Meyer ([**CoM**]).

Discrete variants of all these 'continuous' examples can be derived by replacing the integrals with approximating Riemann sums. Such Cotlar type operators as wavelet and frame operators arise in this way from the homogeneous case $\mathcal{C}_{\phi\psi}$. For each fixed (r, s), $r > 1, s > 0$ choose an affine lattice

$$(2.14)(\text{i}) \qquad \lambda_{j,k} = (sk/r^j, 1/r^j) \qquad (k \in \mathbb{Z}^n, j \in \mathbb{Z})$$

in H and let Δ be the *fundamental tile*

$$(2.14)(\text{ii}) \qquad \Delta = \{((y_1, \ldots, y_n), t) \in H : 0 \le y_\ell < s \,, 1 \le t < r\}$$

in H. Then the union $\bigcup_{j,k} \Delta.\lambda_{j,k}$ of right translates of Δ provides an affine tiling of H by tiles all having the same measure, *i.e.*, $|\Delta.\lambda_{j,k}| = |\Delta|$ independently of j and k because of the use of right invariant Haar measure. Now define a, b by

$$(2.15)(\text{i}) \qquad a(z\,;y) = \frac{1}{|\Delta|^{1/2}} \sum_{j,k} \chi_\Delta(z\lambda_{j,k}^{-1})U(z\lambda_{j,k}^{-1})\phi(y)$$

and

$$(2.15)(\text{ii}) \qquad b(z\,;x) = \frac{1}{|\Delta|^{1/2}} \sum_{j,k} \chi_\Delta(z\lambda_{j,k}^{-1})U(z\lambda_{j,k}^{-1})\psi(x)$$

where χ_Δ is the characteristic function of Δ. By construction

$$\|a(z\,;.)\|_{\mathcal{M}_\delta} \le \frac{1}{|\Delta|^{1/2}} \sup_{w \in \Delta} \|U(w)\phi\|_{\mathcal{M}_\delta} \le \text{const.} \|\phi\|_{\mathcal{M}_\delta}$$

since the union of the $\Delta.\lambda_{j,k}$ provides an affine tiling of H and $U(w)$ is bounded on compact sets of H with constant depending only on Δ. There is a similar estimate for $b(z\,;\,x)$. Thus a,b are L^∞-bounded functions on H and

$$U^*(z)b(z;x)\overline{U^*(z)a(z;y)}$$

$$= \frac{1}{|\Delta|}\sum_{j,k}\chi_\Delta(z\lambda_{j,k}^{-1})U^*(\lambda_{j,k})\psi(x)\overline{U^*(\lambda_{j,k})\phi(y)}$$

$$= \frac{1}{|\Delta|}\sum_{j,k}r^{jn}\chi_\Delta(z\lambda_{j,k}^{-1})\psi(r^j x - sk)\overline{\phi(r^j y - sk)}.$$

The corresponding $\mathcal{T}_{ab}$ is just the frame operator

$$\mathcal{D}_{\phi\psi} : f \longrightarrow \sum_{j,k}\langle f,\, U^*(\lambda_{j,k})\phi\rangle U^*(\lambda_{j,k})\psi$$

$$= \sum_{j,k}r^{jn}\left(\int_{\mathbb{R}^n} f(y)\overline{\phi(r^j y - sk)}\,dy\right)\psi(r^j x - sk)$$

$$= \sum_{j,k}\langle f,\phi_{j,k}\rangle\psi_{j,k}$$

associated with ϕ,ψ. It is natural to call this a *Discrete Sum Operator* or, more precisely, a *Discrete Sum Operator having mesh size* (r,s) to emphasize its dependence on the affine lattice $\{\lambda_{j,k}\}$. Obviously $\mathcal{D}_{\phi\psi}$ has the bilateral vanishing moment property if both of ϕ,ψ have vanishing moment; on the other hand, if only one of ϕ or ψ has vanishing moment $\mathcal{D}_{\phi\psi}$ will have the unilateral vanishing moment property. The prototypical example is the wavelet expansion $f = \mathcal{D}_{\psi\psi}f$ which occurs when ϕ,ψ are the same mother wavelet and the tiling is the usual dyadic tiling of $\mathbb{R}^2_+$; here $\mathcal{D}_{\phi\psi}$ will always have the bilateral vanishing moment property. More generally, these discrete sum operators can be used to study frames and frame decompositions.

The final example arises in a very different context — Uchiyama's constructive decomposition of BMO-functions ([**GHL**], [**Uch**]). We start with the Discrete sum operator

$$\mathcal{D}_{\psi\psi} : f \longrightarrow \sum_{j,k}\langle f,\, U^*(\lambda_{j,k})\psi\rangle U^*(\lambda_{j,k})\psi.$$

In the constructive decomposition described in [**GHL**] these functions are mother wavelets, in fact, but in Uchiyama's prescription they were based on the notion of 'elementary particles' introduced by R. Fefferman and Chang. Now suppose that to each pair (j,k) there is given a standard singular integral operator $L_{j,k}$, *i.e.*, commuting with translation and dilation, whose standard estimates are uniformly bounded in j and k. Such $L_{j,k}$ arise at a critical stage during the course of the constructive decomposition. Then define $\mathcal{L}$ by

$$\mathcal{L} : f \longrightarrow \sum_{j,k}\langle f,\, U^*(\lambda_{j,k})\psi\rangle L_{j,k}(U^*(\lambda_{j,k})\psi)$$

$$= \sum_{j,k}\langle f,\, U^*(\lambda_{j,k})\psi\rangle U^*(\lambda_{j,k})\Psi_{j,k}$$

where $\Psi_{j,k} = L_{j,k}\psi$. This is not of Discrete sum type because the synthesizing functions are not translates and dilates of a *fixed* function but it still is a $\mathcal{T}_{ab}$ operator for appropriate choices of a, b.

3. Kernel estimates

In this section we show that the distributional kernel of a $\mathcal{T}_{ab}$-operator satisfies all the estimates in the previous section with only minimal loss of smoothness. Fix ε, $0 < \varepsilon < \delta - [\delta]$.

(3.1) THEOREM. *Let $K(x,y)$ be the kernel (2.10) associated with a pair of L^∞-bounded functions $a, b : H \to \mathcal{M}_\delta(\mathbb{R}^n)$. Then for $(x,y) \in \Omega$ the inequalities*

(i) $$|D_x^\alpha D_y^\beta K(x,y)| \;\leq\; \text{const.}\,\frac{C_{ab}}{|x-y|^{n+|\alpha|+|\beta|}}\;,$$

(ii) $$|D_x^\alpha D_y^\nu K(x,y) - D_x^\alpha D_y^\nu K(x',y)| \;\leq\; \text{const.}\,C_{ab}\,\frac{|x-x'|^\varepsilon}{|x-y|^{n+|\alpha|+[\delta]+\varepsilon}}$$

(iii) $$|D_x^\nu D_y^\beta K(x,y) - D_x^\nu D_y^\beta K(x,y')| \;\leq\; \text{const.}\,C_{ab}\,\frac{|y-y'|^\varepsilon}{|x-y|^{n+|\beta|+[\delta]+\varepsilon}}$$

hold for all $|\alpha|, |\beta| \leq [\delta]$, $|\nu| = [\delta]$ and $|x - x'|, |y - y'| \leq \frac{1}{2}|x - y|$ where $C_{ab} = \|a\|_{\infty,\delta}\,\|b\|_{\infty,\delta}$.

PROOF OF THEOREM (3.1). In view of A(i),

$$|K(x,y)| \;\leq\; \text{const.}\,C_{ab}\int_0^\infty \frac{t^\delta}{(t+|x-y|)^{n+\delta}}\,\frac{dt}{t} \;\leq\; \text{const.}\,\frac{C_{ab}}{|x-y|^n}$$

uniformly in a, b whenever $x \neq y$. All the remaining inequalities require smoothness as well as decay. Since this smoothness need not be the same for each variable, an alternative to A(i) is needed.

BASIC ESTIMATE A(ii). *The inequality*

$$\int_0^\infty \int_{\mathbb{R}^n} \frac{t^\delta}{(t+|x-v|)^{n+\sigma}}\,\frac{t^\delta}{(t+|v|)^{n+\tau}}\,\frac{dv\,dt}{t} \;\leq\; \text{const.}\,|x|^{2\delta-n-\sigma-\tau}$$

holds uniformly in x provided σ, τ are positive and $\max(\sigma,\tau) < 2\delta < \sigma + \tau$.

Again a proof appears in the Appendix.

Recall that we can write the kernel as

$$D_x^\alpha D_y^\beta K(x,y) \;=\; \int_0^\infty \frac{1}{t^{2n}}\left(\int_{\mathbb{R}^n} D_x^\alpha b(z;\, x.z^{-1})\,\overline{D_y^\beta(z;\, y.z^{-1})}\,dv\right)\frac{dt}{t}$$

and so

$$|D_x^\alpha D_y^\beta K(x,y)| \;\leq\; C_{ab}\int_0^\infty\left(\int_{\mathbb{R}^n} \frac{t^\delta}{(t+|y-v|)^{n+\delta+|\alpha|}}\,\frac{t^\delta}{(t+|x-v|)^{n+\delta+|\beta|}}\,dv\right)\frac{dt}{t}.$$

Inequality (3.1)(i) follows immediately using A(ii).

It is in the corresponding Lipschitz estimates that the loss of smoothness becomes apparent. We first remark that

$$\left|(D_x^\alpha b)(z;x) - (D_x^\alpha b)(z;x)\right|$$

$$\leq\; \text{const.}\,\|b\|_{\infty,\delta}|x-x'|^\varepsilon\left\{\frac{1}{(1+|x|)^{n+\delta+[\delta]+\varepsilon}} + \frac{1}{(1+|x'|)^{n+\delta+[\delta]+\varepsilon}}\right\}$$

for $|\alpha| = [\delta]$, $0 < \varepsilon \le \delta - [\delta]$ and all $x, x' \in \mathbb{R}^n$. See lemma (2.5) in chapter 1. Thus

$$|D_x^\alpha D_y^\beta K(x, y) - D_x^\alpha D_y^\beta K(x', y)|$$

$$\le \int_H \frac{1}{t^{2n}} \left| D_x^\alpha b\left(z; \frac{x - v}{t}\right) - D_x^\alpha b\left(z; \frac{x' - v}{t}\right) \right| \left| D_y^\beta a\left(z; \frac{y - v}{t}\right) \right| dv \frac{dt}{t}$$

$$\le \text{const.} \, C_{ab} |x - x'|^\varepsilon \int_H \frac{t^\delta}{(t + |x - v|)^{n+\delta+[\delta]+\varepsilon}} \frac{t^\delta}{(t + |y - v|)^{n+\delta+|\beta|}} dv \frac{dt}{t}$$

$$+ \text{ const.} \, C_{ab} |x - x'|^\varepsilon \int_H \frac{t^\delta}{(t + |x' - v|)^{n+\delta+[\delta]+\varepsilon}} \frac{t^\delta}{(t + |y - v|)^{n+\delta+|\beta|}} dv \frac{dt}{t}$$

$$\le \text{const.} \, C_{ab} |x - x'|^\varepsilon \left(|x - y|^{-n-[\delta]-\delta-\varepsilon} + |x' - y|^{-n-[\delta]-\delta-\varepsilon} \right)$$

by basic estimate A(ii) so long as $0 < \varepsilon < \delta - [\delta]$. For $|x - x'| \le \frac{1}{2}|x - y|$ we have $|x - y| \sim |x' - y|$. Therefore

$$\left| D_x^\alpha D_y^\beta K(x, y) - D_x^\alpha D_y^\beta K(x', y) \right| \le \text{const.} \, C_{ab} |x - x'|^\varepsilon |x - y|^{-n-[\delta]-|\beta|-\varepsilon}$$

for $|x - x'| \le \frac{1}{2}|x - y|$. $\qquad\qquad\qquad\qquad\qquad\qquad\qquad\qquad\qquad\qquad\Box$

Much the same proof establishes also the joint Lipschitz condition for $K(x, y)$.

(3.2) THEOREM. *The kernel $K(x, y)$ also satisfies the joint Lipschitz condition*

$$\left| [D_x^\alpha D_y^\beta K(x, y) - D_x^\alpha D_y^\beta K(x', y)] - [D_x^\alpha D_y^\beta K(x, y') - D_x^\alpha D_y^\beta K(x', y')] \right|$$

$$\le \text{ const.} \, \|a\|_{\delta,\infty} \|b\|_{\delta,\infty} \frac{|x - x'|^\varepsilon |y - y'|^\varepsilon}{|x - y|^{n+2[\delta]+2\varepsilon}} \qquad (|\alpha|, |\beta| = [\delta])$$

for $0 < \varepsilon < \delta - [\delta]$ whenever $|x - x'|, |y - y'| \le \frac{1}{3}|x - y|$.

PROOF OF THEOREM (3.2). For $0 < \varepsilon < \delta - [\delta]$ we have

$$\left| [D_x^\alpha D_y^\beta K(x, y) - D_x^\alpha D_y^\beta K(x', y)] - [D_x^\alpha D_y^\beta K(x, y') - D_x^\alpha D_y^\beta K(x', y')] \right|$$

$$\le \text{const.} \, C_{ab} |x - x'|^\varepsilon |y - y'|^\varepsilon \int_H \left| \frac{t^\delta}{(t + |x - v|)^{n+\delta+[\delta]+\varepsilon}} + \frac{t^\delta}{(t + |x' - v|)^{n+\delta+[\delta]+\varepsilon}} \right|$$

$$\times \left| \frac{t^\delta}{(t + |y - v|)^{n+\delta+[\delta]+\varepsilon}} + \frac{t^\delta}{(t + |y' - v|)^{n+\delta+[\delta]+\varepsilon}} \right| dv \frac{dt}{t}$$

$$= I + II + III + IV$$

where we have split the absolute value signs to obtain four terms. The first one we estimate as

$$I \le \text{const.} \, C_{ab} |x - x'|^\varepsilon |y - y'|^\varepsilon \int \frac{t^\delta}{(t + |x - v|)^{n+\delta+[\delta]+\varepsilon}}$$

$$\times \frac{t^\delta}{(t + |y - v|)^{n+\delta+[\delta]+\varepsilon}} dv \frac{dt}{t}$$

$$\le \text{const.} \, C_{ab} |x - x'|^\varepsilon |y - y'|^\varepsilon |x - y|^{-n-2[\delta]-2\varepsilon}.$$

For $|x - x'|, |y - y'| \le \frac{1}{3}|x - y|$ we see that $|x - y| \sim |x' - y| \sim |x - y'| \sim |x' - y'|$. Thus

$$\left| [D_x^\alpha D_y^\beta K(x,y) - D_x^\alpha D_y^\beta K(x',y)] - [D_x^\alpha D_y^\beta K(x,y') - D_x^\alpha D_y^\beta K(x',y')] \right|$$

$$\le \text{ const. } C_{ab}|x - x'|^\varepsilon |y - y'|^\varepsilon \Big(|x - y|^{-n-2[\delta]-2\varepsilon} + |x' - y|^{-n-2[\delta]-2\varepsilon}$$

$$+ |x - y'|^{-n-2[\delta]-2\varepsilon} + |x' - y'|^{-n-2[\delta]-2\varepsilon} \Big)$$

$$\le \text{ const. } C_{ab}|x - x'|^\varepsilon |y - y'|^\varepsilon |x - y|^{-n-2[\delta]-2\varepsilon}$$

for $|x - x'|, |y - y'| \le \frac{1}{3}|x - y|$. $\qquad\square$

In the proofs above, notice that it is the restriction $\max(\sigma, \tau) < 2\delta$ in A(ii) that forces the loss of smoothness in the kernel. We can now apply these results to $\mathcal{T}_{ab}$-operators. The first result follows at once from (3.1) and definition (1.2).

(3.3) THEOREM. *Let $\mathcal{T}_{ab}$ be an operator of Cotlar type associated with L^∞-bounded functions $a, b : H \to \mathcal{M}_\delta(\mathbb{R}^n)$. Then $\mathcal{T}_{ab}$ belongs to $CZO(\gamma)$ for each γ, $\gamma < \delta$.*

Now suppose that a has vanishing moments. Then $\mathcal{T}_{ab}$ is of Cotlar type and maps $\mathcal{M}_\delta$ into $L^2(\mathbb{R}^n)$; in particular, $\mathcal{T}_{ab}$ will have the Weak Boundedness Property as well as the Restricted Boundedness Property. Together with (3.2) and (3.3) this shows

(3.4) COROLLARY. *Let $a : H \to \mathcal{M}_\delta^{(0)}(\mathbb{R}^n)$, $b : H \to \mathcal{M}_\delta(\mathbb{R}^n)$ be L^∞-bounded functions on H. Then the associated Cotlar type operator $\mathcal{T}_{ab}$ is a $CZO(\gamma)$-operator satisfying condition $\mathcal{H}(\gamma)$ for which the inequality*

$$\|\mathcal{T}_{ab}\|_{\mathcal{H}} \le \text{ const. } \|a\|_{\infty,\delta} \|b\|_{\infty,\delta}$$

holds uniformly in a, b for each γ, $\gamma < \delta$.

When both a, b have vanishing moments then either form of the $T(\mathbf{1})$-theorem ensures that $\mathcal{T}_{ab}$ is bounded on $L^2(\mathbb{R}^n)$. Indeed, $\mathcal{T}_{ab}$ will have WBP and $\mathcal{T}_{ab}(1) = \mathcal{T}_{ab}^*(1) = 0$, so the David-Journé version applies. On the other hand, the inequalities

$$\|U^*(z)\mathcal{T}_{ab}U(z)f\|_{L^2}, \quad \|U^*(z)\mathcal{T}_{ab}^*U(z)f\|_{L^2} \le \text{ const. } C_{ab}\|f\|_{\mathcal{M}_\delta}$$

hold uniformly in z when a and b have vanishing moments, so Stein's version applies also. Either thus establishes the following result.

(3.5) THEOREM. *Let $a, b : H \to \mathcal{M}_\delta^{(0)}(\mathbb{R}^n)$ be L^∞-bounded functions on H and let $\mathcal{T}_{ab}$ be the associated operator of Cotlar type. Then $\mathcal{T}_{ab}$ is bounded on $L^2(\mathbb{R}^n)$; furthermore, the inequality*

$$\|T_{ab}f\|_{L^2} \le \text{ const. } C_{ab}\|f\|_{L^2} \qquad (f \in L^2)$$

then holds uniformly in a, b and f.

Although strictly speaking it is not needed in this paper, there is a natural extension of (3.5) which applies to examples which need not satisfy the bilateral vanishing moment condition. One 'obvious' way of showing that $\mathcal{T}_{ab}$ is bounded from $L^2(\mathbb{R}^n)$ into $L^2(\mathbb{R}^n)$ is to express T_{ab} as the composition $\mathfrak{M}_b^* \circ \mathfrak{M}_a$ of the 'matrix coefficient' mappings

$$\mathfrak{M}_a : f \longrightarrow \langle f, U^*(z)a(z) \rangle = \frac{1}{t^{n/2}} \int_{\mathbb{R}^n} \overline{a(z\,;\,y.z^{-1})} f(y)\,dy.$$

Each $\mathcal{M}_a$ maps functions on $\mathbb{R}^n$ to functions on H and so the L^2-boundedness will follow immediately once it is known that both $\mathcal{M}_a$ and $\mathcal{M}_b$ map $L^2(\mathbb{R}^n)$ into $L^2(H)$. In the homogeneous case when a, b are independent of z, each $\mathfrak{M}_a$ is a matrix coefficient mapping in the traditional sense for the representation $z \to U(z)$ of H and the boundedness from $L^2(\mathbb{R}^n)$ into $L^2(H)$ expresses the square integrability of this representation. The precise condition for boundedness in general was determined by Christ and Journé ([**ChJ**]). Recall that a 'tent' T_B in the upper half-space $\mathbb{R}^{n+1}_+$ with base $B = \{v : |v - u| \le s\}$ is the set $T_B = \{(v, t) : |v - u| \le s - t\}$ and that a function $F = F(v, t)$ on $\mathbb{R}^n_+$ is said to belong to the 'Tent space' $\mathcal{N}^\infty$ precisely when

$$\|F\|_{\mathcal{N}^\infty} = \sup_B \left(\frac{1}{|B|} \int_{T_B} |F(v, t)|^2 \frac{dv\,dt}{t} \right)^{1/2}$$

is finite. But with respect to group multiplication on $\mathbb{R}^{n+1}_+ \sim H$ and the usual distance $|z| = |(v, t)| = |v| + t$ a tent T_B can be realized as the set

$$T_B = \{z \in H : |zw^{-1}| \le 1\}, \qquad w = (u, s), \ z = (v, t).$$

Consequently, $\mathcal{N}^\infty$ consists of those functions $F = F(z)$ on H for which

$$\|F\|_{\mathcal{N}^\infty} = \sup_{w \in H} \left(\int_{|z| \le 1} |F(zw)|^2 \frac{dv\,dt}{t} \right)^{1/2}$$

is finite; in particular, the right translate $F_\zeta, F_\zeta(z) = F(z\zeta), \zeta \in H$, of any F in $\mathcal{N}^\infty(H)$ is again in $\mathcal{N}^\infty(H)$ and $\|F_\zeta\|_{\mathcal{N}^\infty} = \|F\|_{\mathcal{N}^\infty}$. With this notation Christ and Journé have shown that the inequality

$$\int_0^\infty \int_{\mathbb{R}^n} |\mathfrak{M}_a(f)(v, t)|^2 \frac{dv\,dt}{t^{n+1}} \le \text{const.} \int_{\mathbb{R}^n} |f(y)|^2\,dy$$

holds for all f in $L^2(\mathbb{R}^n)$ if and only if the function $t^{-n/2}\mathfrak{M}_a(\mathbf{1})(z)$ belongs to the tent space $\mathcal{N}^\infty(H)$; furthermore, the inequality

$$(3.6) \qquad \|\mathfrak{M}_a(f)\|_{L^2(H)} \le \text{const.}\{\|a\|_{\infty,\delta} + \|t^{-n/2}\mathfrak{M}_a(\mathbf{1})\|_{\mathcal{N}^\infty}\}\|f\|_{L^2}$$

then holds uniformly in a and f ([**ChJ**]). Since

$$t^{-n/2}\mathfrak{M}_a(\mathbf{1})(z) = \frac{1}{t^n} \int_{\mathbb{R}^n} \overline{a(z\,;\,\frac{y-v}{t})}\,dy = \overline{\int_{\mathbb{R}^n} a(z\,;\,y)\,dy}$$

is just the (complex conjugate) of the *moment* of a, we obtain a natural extension of theorems (2.12) and (3.5).

(3.7) THEOREM. *Let $a, b : H \to \mathcal{M}_\delta(\mathbb{R}^n)$ be L^∞-bounded functions whose moments*

$$A(z) = \int_{\mathbb{R}^n} a(z\,;y)\,dy, \qquad B(z) = \int_{\mathbb{R}^n} b(z\,;x)\,dx$$

define functions in the Tent space $\mathcal{N}^\infty(H)$. Then the associated $\mathcal{T}_{ab}$ operator is of Cotlar type. Furthermore, $\mathcal{T}_{ab}$ defines a bounded operator on $L^2(\mathbb{R}^n)$ such that

$$\|\mathcal{T}_{ab}(f)\|_{L^2} \leq \text{const.}(\|a\|_{\infty,\delta} + \|A\|_{\mathcal{N}^\infty})(\|b\|_{\infty,\delta} + \|B\|_{\mathcal{N}^\infty})\|f\|_{L^2}$$

uniformly in a, b and f.

While theorem (3.5) would seem to be the trivial case $A = B = 0$ of (3.7), it should be noted that all such results are based in the final analysis on Cotlar's lemma in the vanishing moment case together with para-product arguments to take care of non-vanishing moments.

PROOF OF THEOREM (3.7). In view of (3.6)

$$\int_0^\infty \int_{\mathbb{R}^n} |\mathfrak{M}_a(f)(v,t)|\, |\mathfrak{M}_b(g)(v,t)|\, \frac{dv\,dt}{t^{n+1}}$$

$$\leq \text{const.}\big(\|a\|_{\infty,\delta} + \|A\|_{\mathcal{N}^\infty}\big)\big(\|b\|_{\infty,\delta} + \|B\|_{\mathcal{N}^\infty}\big)\|f\|_{L^2}\|g\|_{L^2}\ .$$

On the other hand,

$$\int_\varepsilon^\infty \left(\int_{\mathbb{R}^n}\int_{\mathbb{R}^n} K_t(x,y)f(y)\overline{g(x)}\,dy dx\right)\frac{dt}{t}$$

$$= \int_\varepsilon^\infty \int_{\mathbb{R}^n} \mathfrak{M}_a(f)(v,t)\,\overline{\mathfrak{M}_b(g)(v,t)}\,\frac{dv\,dt}{t^{n+1}}.$$

Thus the limit in (2.9) exists and defines a bounded operator from $L^2(\mathbb{R}^n)$ into $L^2(\mathbb{R}^n)$ whose norm is as stated. $\qquad\square$

4. Boundedness on spaces of molecules

Finally we come to the principal results of this chapter. The first generalizes Han's result to arbitrary $\delta > 0$, while the second uses the generalization to exhibit what for practical purposes is a characterization of the class of operators satisfying condition $\mathcal{H}$. The generalization to be proved is thus the one given earlier as theorem (1.8).

(4.1) THEOREM. *Let T be a $CZO(\delta)$-operator satisfying condition $\mathcal{H}(\delta)$. Then for each fixed fixed γ, $\gamma < \delta$, T maps any $\mathcal{M}_\gamma$-molecule into an $\mathcal{M}_\gamma$-test function preserving center and width; more precisely, the inequality*

$$(*) \qquad \|U^*(z)TU(z)(\psi)\|_{\mathcal{M}_\gamma} \leq \text{const.}\,\|T\|_{\mathcal{H}}\,\|\psi\|_{\mathcal{M}_\gamma} \qquad (z \in H)$$

holds uniformly in T and z for all $\mathcal{M}_\gamma$-molecules ψ.

In view of corollary (3.4), therefore, theorem (4.1) applies to any $\mathcal{T}_{ab}$-operator in which a has vanishing moment. In fact, these are essentially the only possible operators.

(4.2) THEOREM. *Let T be a $CZO(\delta)$-operator that satisfies condition $\mathcal{H}(\delta)$. Then for each γ, $\gamma < \delta$, there exist L^∞-bounded functions*

$$a : H \to \mathcal{M}_\gamma^{(0)}(\mathbb{R}^n), \qquad b : H \to \mathcal{M}_\gamma(\mathbb{R}^n),$$

so that $T = \mathcal{T}_{ab}$.

Let us postpone the proof of (4.1) for the moment and proceed immediately to (4.2) since the latter brings out very clearly how both Cotlar type representations and wavelet decompositions can be used to provide molecular decompositions for $CZO(\delta)$-operators much as the 'non-standard' decomposition of operators does.

PROOF OF THEOREM (4.2). To simplify the proof, suppose $n = 1$ and let ψ be a compactly supported mother wavelet having both vanishing moments and continuous derivatives of order at least $[\delta] + 1$. We shall use the construction of dyadic Discrete sum examples of $\mathcal{T}_{ab}$-operators given in section 2, *i.e.*, $r = 2, s = 1$. Each f in, say, $\mathcal{M}_\delta(\mathbb{R})$ has an unconditionally convergent wavelet expansion $f = \sum_{j,k}\langle f, \psi_{j,k}\rangle\psi_{j,k}$ with respect to the dyadic translates and dilates

$$\psi_{j,k}(x) = 2^{j/2}\psi(2^j x - k) = U(\lambda_{j,k}^{-1})\psi, \qquad \lambda_{j,k} = (k/2^j, 1/2^j)$$

of ψ. Now set $\phi^{(j,k)} = U(\lambda_{j,k})TU(\lambda_{j,k}^{-1})\psi$ and define a, b on H by

$$a(z\,;\,y) = \frac{1}{|\Delta|^{1/2}}\sum_{j,k}\chi_\Delta(z\lambda_{j,k}^{-1})U(z\lambda_{j,k}^{-1})\psi(y)$$

and

$$b(z\,;\,x) = \frac{1}{|\Delta|^{1/2}}\sum_{j,k}\chi_\Delta(z\lambda_{j,k}^{-1})U(z\lambda_{j,k}^{-1})\phi^{(j,k)}$$

respectively. As in (2.15) a is an L^∞-bounded function with values in $\mathcal{M}_\delta^{(0)}(\mathbb{R})$. On the other hand, by (4.1), $\{\phi^{(j,k)} : j, k \in \mathbb{Z}\}$ is a uniformly norm-bounded family of functions in $\mathcal{M}_\delta(\mathbb{R})$. Thus b will be an L^∞-bounded function taking values in $\mathcal{M}_\delta(\mathbb{R})$. Furthermore, $Tf = \mathcal{T}_{ab}f$, completing the proof in case $n = 1$. This same proof in higher dimensions will yield a sum of $\mathcal{T}_{ab}$ operators because of the need for $2^n - 1$ mother wavelets. To avoid this one can use the $\varphi - \psi$-transform of Frazier and Jawerth ([**FJ**]). $\qquad\square$

The same proof as for (4.2) shows that condition $\mathcal{H}$ is also necessary for a $CZO(\delta)$-operator to preserve smoothness as well as the width and center of molecules. To be precise

(4.3) THEOREM. *Suppose $T : \mathcal{M}_\gamma^{(0)}(x_0, d) \to \mathcal{M}_\gamma(x_0, d)$ uniformly in $x_0 \in \mathbb{R}^n$ and $d > 0$. Then for each $\gamma' < \gamma$, $T \in CZO(\gamma')$ and T satisfies condition $\mathcal{H}(\gamma')$.*

PROOF OF THEOREM (4.3). As the proof of (4.2) shows, any such T can be written as

$$T = \mathcal{T}_{ab}, \qquad a : H \to \mathcal{M}_\gamma^{(0)}(\mathbb{R}^n), \qquad b : H \to \mathcal{M}_\gamma(\mathbb{R}^n)\,.$$

The theorem now follows immediately from corollary (3.4). $\qquad\square$

PROOF OF THEOREM (4.1). Since the operator $U^*(z)TU(z)$ satisfies the hypotheses of the theorem uniformly in $z \in H$ it suffices to demonstrate that T is bounded on the space of $\mathcal{M}_\gamma$-molecules centered at the origin with width 1. Fix $\varphi \in \mathcal{M}_\gamma^{(0)} \cap \mathcal{D}$. We need to show that $T\varphi$ is an $\mathcal{M}_\gamma$-test function. This involves showing that

$$(*) \qquad |D^\beta(T\varphi)(x)| \leq \text{const.}\,\|\varphi\|_{\mathcal{M}_\gamma}(1 + |x|)^{-n-\gamma-|\beta|}$$

for $|\beta| \leq [\gamma]$ and

$$(**) \quad |D^\beta(T\varphi)(x) - D^\beta(T\varphi)(x')| \leq \text{const.} \|\varphi\|_{\mathcal{M}_\gamma} |x - x'|^{\gamma - [\gamma]} (1 + |x|)^{-n-2\gamma}$$

for $|\beta| = [\gamma]$ and $|x - x'| \leq \frac{1}{2}(1 + |x|)$.

We first need a technical lemma. For a continuous linear operator $T : \mathcal{D} \to \mathcal{D}'$ with distribution kernel K, and two multi-indices α and β, we denote by $T_{\alpha,\beta}$ the continuous linear operator from $\mathcal{D}$ to $\mathcal{D}'$ whose distribution kernel is

$$K_{\alpha,\beta} = (y - x)^\alpha D_x^\beta K .$$

The following lemma is due to R.H. Torres ([**Tor**]).

(4.4) LEMMA. *Let* $T \in CZO(\delta) \cap WBP$ *and let* $|\alpha| \leq |\beta| \leq [\delta]$. *If* $f \in \mathcal{D}(\mathbb{R}^n \times \mathbb{R}^n)$ *and* $D_x^\lambda D_y^\mu f(x, x) = 0$ *for all* $|\lambda| + |\mu| \leq |\beta| - |\alpha|$, *then*

$$\langle K_{\alpha,\beta}, f \rangle = \int\int_{x \neq y} (y - x)^\alpha D_x^\beta K(x, y) f(x, y) \, dxdy ,$$

where the integral on the right converges absolutely.

Fix $\theta \in \mathcal{D}$ such that $\text{supp}\,\theta \subset \{|x| \leq 4\}$ and $\theta \equiv 1$ on $\{|x| \leq 2\}$. Define η so that $1 = \theta + \eta$.

STEP 1. We prove $(*)$ for $|x| \leq 1$. Assume for simplicity that $|\beta| = [\gamma]$. Using (4.4) we can write

$$D^\beta(T\varphi)(x) = \int D_x^\beta K(x, y) \Big[\varphi(y) - \sum_{|\alpha| \leq [\gamma]} \frac{D^\alpha \varphi(x)}{\alpha!} (y - x)^\alpha \Big] \theta(y) \, dy$$

$$+ \int D_x^\beta K(x, y)\, \varphi(y)\, \eta(y) \, dy$$

$$+ \int D_x^\beta K(x, y) \sum_{|\alpha| \leq [\gamma]} \frac{D^\alpha \varphi(x)}{\alpha!} (y - x)^\alpha \theta(y) \, dy$$

$$= I + II + III.$$

To estimate the Taylor series remainder we use the mean value theorem.

$$\varphi(y) - \sum_{|\alpha| \leq [\gamma]} \frac{D^\alpha \varphi(x)}{\alpha!} (y - x)^\alpha$$

$$= \varphi(y) - \sum_{|\alpha| < [\gamma]} \frac{D^\alpha \varphi(x)}{\alpha!} (y - x)^\alpha - \sum_{|\alpha| = [\gamma]} \frac{D^\alpha \varphi(x)}{\alpha!} (y - x)^\alpha$$

$$= \sum_{|\alpha| = [\gamma]} \frac{D^\alpha \varphi(z) - D^\alpha \varphi(x)}{\alpha!} (y - x)^\alpha$$

where $|z - x| < |y - x|$. Since $\varphi \in \mathcal{M}_\gamma^{(0)}$ we have

$$\Big| \varphi(y) - \sum_{|\alpha| \leq [\gamma]} \frac{D^\alpha \varphi(x)}{\alpha!} (y - x)^\alpha \Big| \leq \text{const.} \|\varphi\|_{\mathcal{M}_\gamma} |x - y|^\gamma (1 + |x|)^{-n-2\gamma}$$

for $|x - y| \leq \frac{1}{2}(1 + |x|)$. Therefore

$$|I| \;\leq\; \text{const.} \|\varphi\|_{\mathcal{M}_\gamma} \int_{|y| \leq 4} |x - y|^{-n - [\gamma]} |x - y|^\gamma \, dy$$

$$\leq \; \text{const.} \|\varphi\|_{\mathcal{M}_\gamma} \int_0^5 r^{-n - [\gamma] + \gamma} r^{n-1} \, dr \;\leq\; \text{const.} \|\varphi\|_{\mathcal{M}_\gamma}$$

since $|x - y| \leq |x| + |y| \leq 5$. For $y \in \operatorname{supp} \eta$ and $|x| \leq 1$ we have $|y| \leq |x| + |x - y| \leq \frac{|y|}{2} + |x - y|$. Thus $\frac{|y|}{2} \leq |x - y|$ and hence

$$|II| \;\leq\; \text{const.} \|\varphi\|_{\mathcal{M}_\gamma} \int_{|y| > 2} |x - y|^{-n - [\gamma]} |y|^{-n - \gamma} \, dy$$

$$\leq \; \text{const.} \|\varphi\|_{\mathcal{M}_\gamma} \int_{|y| > 2} |y|^{-n - [\gamma]} |y|^{-n - \gamma} \, dy \;\leq\; \text{const.} \|\varphi\|_{\mathcal{M}_\gamma}.$$

To estimate III we need another lemma, also due to R.H. Torres ([**Tor**]).

(4.5) LEMMA. *Suppose* $T \in CZO(\delta) \cap WBP$ *and* $Ty^\mu = 0$ *for all* $|\mu| \leq [\delta]$. *Let* $|\alpha| \leq |\beta| \leq [\delta]$. *Then* $T_{\alpha,\beta}$ *maps* $\mathcal{D}$ *into* L^∞. *Moreover,*

$$\left\| T_{\alpha,\beta} \psi \left(\frac{\cdot - z}{t} \right) \right\|_\infty \;\leq\; \text{const.} \, t^{|\alpha| - |\beta|}$$

where the constant only depends on T *and* ψ.

Finally, we can apply this lemma to obtain

$$|III| \;\leq\; \sum_{|\alpha| \leq [\gamma]} \frac{|D^\alpha \varphi(x)|}{\alpha!} \left| \int (y - x)^\alpha D_x^\beta K(x, y) \, \theta(y) \, dy \right|$$

$$\leq \; \text{const.} \sum_{|\alpha| \leq [\gamma]} |D^\alpha \varphi(x)| \, |T_{\alpha,\beta} \theta(x)|$$

$$\leq \; \text{const.} \sum_{|\alpha| \leq [\gamma]} \|D^\alpha \varphi\|_\infty \|T_{\alpha,\beta} \theta\|_\infty$$

$$\leq \; \text{const.} \|\varphi\|_{\mathcal{M}_\gamma}.$$

STEP 2. We now show that $(*)$ holds for $|x| > 1$. With θ as before, set $1 = I + J + L$ where

$$I(y) \;=\; \theta\left(\frac{8(x - y)}{|x|} \right) \qquad J(y) \;=\; \theta\left(\frac{8y}{|x|} \right).$$

Now set $\varphi = \varphi_I + \varphi_J + \varphi_L = \varphi \cdot I + \varphi \cdot J + \varphi \cdot L$. We will estimate these functions separately. First,

$$|D^\beta(T\varphi_I)(x)| \leq \left| \int D_x^\beta K(x,y) \big[\varphi(y) - \sum_{|\alpha| \leq [\gamma]} \frac{D^\alpha \varphi(x)}{\alpha!}(y-x)^\alpha\big] I(y)\, dy \right|$$

$$+ \left| \int D_x^\beta K(x,y) \sum_{|\alpha| \leq [\gamma]} \frac{D^\alpha \varphi(x)}{\alpha!}(y-x)^\alpha I(y)\, dy \right|$$

$$= I + II.$$

Using our estimate on the Taylor remainder formula

$$I \leq \text{const.} \|\varphi\|_{\mathcal{M}_\gamma} \int_{|x-y| \leq \frac{|x|}{2}} |x-y|^{-n-[\gamma]} |x-y|^\gamma (1+|x|)^{-n-2\gamma}\, dy$$

$$\leq \text{const.} \|\varphi\|_{\mathcal{M}_\gamma} (1+|x|)^{-n-2\gamma} \int_0^{\frac{|x|}{2}} r^{-n-[\gamma]+\gamma} r^{n-1}\, dr$$

$$\leq \text{const.} \|\varphi\|_{\mathcal{M}_\gamma} (1+|x|)^{-n-\gamma-[\gamma]} .$$

Using lemma (4.5) we obtain

$$II \leq \text{const.} \sum_{|\alpha| \leq [\gamma]} |D^\alpha \varphi(x)| \left\| T_{\alpha,\beta}\theta\left(\frac{8(x-\cdot)}{|x|}\right) \right\|_\infty$$

$$\leq \text{const.} \|\varphi\|_{\mathcal{M}_\gamma} \sum_{|\alpha| \leq [\gamma]} (1+|x|)^{-n-\gamma-|\alpha|} |x|^{|\alpha|-|\beta|}$$

$$\leq \text{const.} \|\varphi\|_{\mathcal{M}_\gamma} (1+|x|)^{-n-\gamma-[\gamma]}.$$

Now we estimate

$$|D^\beta(T\varphi_J)(x)| \leq \left| \int \big[D_x^\beta K(x,y) - \sum_{|\alpha| \leq [\gamma]} \frac{D_y^\alpha D_x^\beta K(x,0)}{\alpha!} y^\alpha\big] \varphi_J(y)\, dy \right|$$

$$+ \left| \sum_{|\alpha| \leq [\gamma]} \frac{D_y^\alpha D_x^\beta K(x,0)}{\alpha!} \int y^\alpha \varphi_J(y)\, dy \right|$$

$$= I + II .$$

Using the estimates on K and the mean value theorem we obtain

$$I \leq \int \left| D^\beta K(x,y) - \sum_{|\alpha| \leq [\gamma]} \frac{D_y^\alpha D_x^\beta K(x,0)}{\alpha!} y^\alpha \right| |\varphi_J(y)| \, dy$$

$$\leq \int \sum_{|\alpha| = [\gamma]} \frac{1}{\alpha!} \left| D_y^\alpha D_x^\beta K(x,z) - D_y^\alpha D_x^\beta K(x,0) \right| |y|^{[\gamma]} |\varphi_J(y)| \, dy$$

$$\leq \text{const.} \|\varphi\|_{\mathcal{M}_\gamma} \int_{|y| \leq \frac{|x|}{2}} |y|^{\delta - [\delta]} |x|^{-n-\delta-[\delta]} |y|^{[\gamma]} |y|^{-n-\gamma} \, dy$$

$$\leq \text{const.} \|\varphi\|_{\mathcal{M}_\gamma} |x|^{-n-\delta-[\delta]} \int_0^{\frac{|x|}{2}} r^{\delta-n-\gamma} r^{n-1} \, dr$$

$$\leq \text{const.} \|\varphi\|_{\mathcal{M}_\gamma} |x|^{-n-\delta-[\delta]} |x|^{\delta-\gamma}$$

$$\leq \text{const.} \|\varphi\|_{\mathcal{M}_\gamma} |x|^{-n-\gamma-[\gamma]} \ .$$

To estimate II we first notice, for $|\alpha| \leq [\gamma]$, that

$$\int |\varphi_I(y)| |y^\alpha| \, dy + \int |\varphi_L(y)| |y^\alpha| \, dy \leq \text{const.} \|\varphi\|_{\mathcal{M}_\gamma} \int_{\frac{|x|}{2}}^\infty r^{-n-\gamma} r^{|\alpha|} r^{n-1} \, dr$$

$$\leq \text{const.} \|\varphi\|_{\mathcal{M}_\gamma} |x|^{|\alpha|-\gamma}$$

since $|y| \geq \frac{|x|}{2}$ for $y \in \operatorname{supp} \varphi_I \cup \operatorname{supp} \varphi_L$. The vanishing moments condition on $\varphi \in \mathcal{M}_\gamma^{(0)}$ implies that

$$\left| \int \varphi_J(y) y^\alpha \, dy \right| = \left| \int \varphi_I(y) y^\alpha \, dy + \int \varphi_L(y) y^\alpha \, dy \right|$$

$$\leq \int |\varphi_I(y)| |y^\alpha| \, dy + \int |\varphi_L(y)| |y^\alpha| \, dy$$

$$\leq \text{const.} \|\varphi\|_{\mathcal{M}_\gamma} |x|^{|\alpha|-\gamma}$$

for $|\alpha| \leq [\gamma]$. Thus

$$II \leq \text{const.} \|\varphi\|_{\mathcal{M}_\gamma} \sum_{|\alpha| \leq [\gamma]} |x|^{-n-|\alpha|-[\gamma]} |x|^{|\alpha|-\gamma}$$

$$\leq \text{const.} \|\varphi\|_{\mathcal{M}_\gamma} |x|^{-n-\gamma-[\gamma]} \ .$$

Finally,

$$|D^\beta(T\varphi_L)(x)| = \left| \int D_x^\beta K(x,y) \varphi_L(y) \, dy \right|$$

$$\leq \text{const.} \|\varphi\|_{\mathcal{M}_\gamma} \int_{|y|, |x-y| \geq \frac{|x|}{4}} |x-y|^{-n-[\gamma]} |y|^{-n-\gamma} \, dy \ .$$

Note, for $|x - y| \geq \frac{|x|}{4}$, that we have $|y| \leq |x| + |x - y| \leq 5|x - y|$. Thus

$$|D^\beta(T\varphi_L)(x)| \leq \text{const.} \|\varphi\|_{\mathcal{M}_\gamma} \int_{|y| > \frac{|x|}{4}} |y|^{-n-[\gamma]}|y|^{-n-\gamma}\, dy$$

$$\leq \text{const.} \|\varphi\|_{\mathcal{M}_\gamma} |x|^{-n-\gamma-[\gamma]} .$$

STEP 3. We can now prove $(**)$ for $|x - x'| \leq 1$ and $|x| \leq 1$. We need one more technical result due to R.H. Torres ([**Tor**]).

(4.6) LEMMA. *Let $T \in CZO(\delta) \cap WBP$ and $Ty^\mu = 0$ for all $|\mu| \leq [\delta]$. Let x and x' be two distinct points of $\mathbb{R}^n$ and let $\tilde{\theta}(y) = \theta(\frac{y-x'}{|x-x'|})$, and $\tilde{\eta} = 1 - \tilde{\theta}$ with θ as before. Then for all $\varphi \in \mathcal{D}$ and $|\beta| \leq [\delta]$*

$$D^\beta(T\varphi)(x) - D^\beta(T\varphi)(x')$$

$$= \int D_x^\beta K(x,y)\Big[\varphi(y) - \sum_{|\alpha| \leq |\beta|} \frac{D^\alpha\varphi(x)}{\alpha!}(y-x)^\alpha\Big]\tilde{\theta}(y)\, dy$$

$$+ \int D_x^\beta K(x',y)\Big[\varphi(y) - \sum_{|\alpha| \leq |\beta|} \frac{D^\alpha\varphi(x')}{\alpha!}(y-x')^\alpha\Big]\tilde{\theta}(y)\, dy$$

$$+ \int [D_x^\beta K(x,y) - D_x^\beta K(x',y)]\Big[\varphi(y) - \sum_{|\alpha| \leq |\beta|} \frac{D^\alpha\varphi(x')}{\alpha!}(y-x')^\alpha\Big]\tilde{\eta}(y)\, dy$$

$$+ \sum_{|\alpha| \leq |\beta|} \frac{1}{\alpha!}\Big[D^\alpha\varphi(x) - \sum_{|\mu| \leq |\beta|-|\alpha|} \frac{D^{\alpha+\mu}\varphi(x')}{\mu!}(x-x')^\mu\Big]T_{\alpha,\beta}\tilde{\theta}(x)$$

where all the integrals are absolutely convergent.

Let I, II, III and IV denote the four terms appearing in this lemma. We first notice that $|\varphi(y) - \sum_{|\alpha| \leq [\gamma]} \frac{D^\alpha\varphi(x)}{\alpha!}(y-x)^\alpha| \leq |y - x|^\gamma$ for all $y \in \mathbb{R}^n$. For $|y - x| \leq 1$ this follows from the Taylor remainder estimate. For $|y - x| > 1$ this is easily estimated, using the mean value theorem, by $\text{const.} \|\varphi\|_{\mathcal{M}_\gamma} |y - x|^{[\gamma]} \leq \text{const.} \|\varphi\|_{\mathcal{M}_\gamma} |y - x|^\gamma$. For $|\beta| = [\gamma]$ we can estimate

$$|I| \leq \int_{|x'-y| \leq 4|x-x'|} |x - y|^{-n-[\gamma]}\Big|\varphi(y) - \sum_{|\alpha| \leq [\gamma]} \frac{D^\alpha\varphi(x)}{\alpha!}(y-x)^\alpha\Big|\, dy$$

$$\leq \text{const.} \|\varphi\|_{\mathcal{M}_\gamma} \int_{|x'-y| \leq 4|x-x'|} |x - y|^{-n-[\gamma]}|y - x|^\gamma\, dy$$

$$\leq \text{const.} \|\varphi\|_{\mathcal{M}_\gamma} \int_0^{5|x-x'|} r^{-n-[\gamma]+\gamma}r^{n-1}\, dr$$

$$\leq \text{const.} \|\varphi\|_{\mathcal{M}_\gamma} |x - x'|^{\gamma-[\gamma]} .$$

II is estimated similarly. For III,

$$|III| \leq \text{const.} \|\varphi\|_{\mathcal{M}_\gamma} \int_{|x'-y|>2|x-x'|} |x-x'|^{\delta-[\delta]} |x'-y|^{-n-\delta} |x'-y|^\gamma \, dy$$

$$\leq \text{const.} \|\varphi\|_{\mathcal{M}_\gamma} |x-x'|^{\delta-[\delta]} \int_{2|x-x'|}^\infty r^{-n-\delta+\gamma} r^{n-1} \, dr$$

$$\leq \text{const.} \|\varphi\|_{\mathcal{M}_\gamma} |x-x'|^{\delta-[\delta]} |x-x'|^{\gamma-\delta}$$

$$\leq \text{const.} \|\varphi\|_{\mathcal{M}_\gamma} |x-x'|^{\gamma-[\gamma]} .$$

Finally,

$$|IV| \leq \sum_{|\alpha|\leq|\beta|} \left| D^\alpha \varphi(x) - \sum_{|\mu|\leq|\beta|-|\alpha|} \frac{D^{\alpha+\mu}\varphi(x')}{\mu!} (x-x')^\mu \right| \left\| T_{\alpha,\beta}\tilde{\theta} \right\|_\infty$$

$$\leq \text{const.} \sum_{|\alpha|+|\mu|\leq|\beta|} \left| D^{\alpha+\mu}\varphi(z) - D^{\alpha+\mu}\varphi(x') \right| |x-x'|^{|\beta|-|\alpha|} |x-x'|^{|\alpha|-|\beta|}$$

$$\leq \text{const.} \|\varphi\|_{\mathcal{M}_\gamma} \sum_{|\alpha|+|\mu|\leq|\beta|} |x-x'|^{\gamma-[\gamma]}$$

$$\leq \text{const.} \|\varphi\|_{\mathcal{M}_\gamma} |x-x'|^{\gamma-[\gamma]} .$$

where $|z-x'| \leq |x-x'|$.

STEP 4. We now prove $(**)$ for $|x| \geq 1$ and $|x-x'| \leq \frac{1}{2}(1+|x|)$. We first estimate $T\varphi_I$. To do this we need the following lemma.

(4.7) LEMMA. *For all $z, w \in \mathbb{R}^n$ and $|\mu| = [\gamma]$ we have*

$$|D^\mu\varphi_I(z) - D^\mu\varphi_I(w)| \leq \text{const.} \|\varphi\|_{\mathcal{M}_\gamma} |z-w|^{\gamma-[\gamma]} |x|^{-n-2\gamma}.$$

PROOF. First suppose that $|z-w| > \frac{|x|}{3}$. If z lies in the support of φ_I then $z \sim x$. Otherwise, $D^\mu\varphi_I(z) = 0$. Similarly for w. Since $|D^\mu\varphi_I(z) - D^\mu\varphi_I(w)| \leq |D^\mu\varphi_I(z)| + |D^\mu\varphi_I(w)|$ we can estimate separately. Each estimate is identical in proof.

$$|D^\mu\varphi_I(z)| \leq \text{const.} \sum_{|\rho|+|\nu|=[\gamma]} |D^\rho\phi(z)\, D^\nu I(z)|$$

$$\leq \text{const.} \|\varphi\|_{\mathcal{M}_\gamma} \sum_{|\rho|+|\nu|=[\gamma]} (1+|z|)^{-n-\gamma-|\rho|} |x|^{-|\nu|}$$

$$\leq \text{const.} \|\varphi\|_{\mathcal{M}_\gamma} \sum_{|\rho|+|\nu|=[\gamma]} \left(\frac{|z-w|}{|x|} \right)^{\gamma-[\gamma]} |x|^{-n-\gamma-|\rho|-|\nu|}$$

$$\leq \text{const.} \|\varphi\|_{\mathcal{M}_\gamma} |z-w|^{\gamma-[\gamma]} |x|^{-n-2\gamma} .$$

On the other hand, suppose $|z - w| \leq \frac{|x|}{3}$. Assume without loss of generality that one of z or w lies in the support of I. Then $|z - x|, |w - x| \leq \frac{5|x|}{6}$. Hence $|z|$ and $|w|$ are comparable to $|x|$. Therefore

$$|D^\mu \varphi_I(z) - D^\mu \varphi_I(w)|$$

$$\leq \text{ const. } \sum_{|\rho|+|\nu|=[\gamma]} |D^\rho \varphi(z)\, D^\nu I(z) - D^\rho \varphi(w)\, D^\nu I(w)|$$

$$\leq \text{ const. } \sum_{|\rho|+|\nu|=[\gamma]} |D^\rho \varphi(z)(D^\nu I(z) - D^\nu I(w))|$$

$$+ \text{ const. } \sum_{|\rho|+|\nu|=[\gamma]} |(D^\rho \varphi(z) - D^\rho \varphi(w))D^\nu I(w)|$$

$$\leq \text{ const. } \|\varphi\|_{\mathcal{M}_\gamma} \sum_{|\rho|+|\nu|=[\gamma]} |x|^{-n-\gamma-|\rho|}|x|^{-|\nu|}$$

$$\times \left[(D^\nu \theta)\left(\frac{8(z-x)}{|x|}\right) - (D^\nu \theta)\left(\frac{8(w-x)}{|x|}\right)\right]$$

$$+ \text{ const. } \|\varphi\|_{\mathcal{M}_\gamma} \sum_{|\rho|+|\nu|=[\gamma]} |x|^{-|\nu|}|z-w|^{\gamma-[\gamma]}|x|^{-n-\gamma-|\rho|-(\gamma-[\gamma])}$$

$$\leq \text{ const. } \|\varphi\|_{\mathcal{M}_\gamma} |z-w|^{\gamma-[\gamma]}|x|^{-n-2\gamma}$$

since $D^\nu \theta$ is certainly Lipschitz of order $\gamma - [\gamma]$. $\qquad\square$

Using lemma (4.6) with the function φ_I and $|\beta| = [\gamma]$ we obtain

$$|D^\beta(T\varphi_I)(x) - D^\beta(T\varphi_I)(x')| \leq I + II + III + IV .$$

To estimate the first term we use lemma (4.7)

$$I \leq \text{ const. } \int_{|x'-y|\leq 4|x-x'|} |x - y|^{-n-[\gamma]}\left| \sum_{|\mu|=[\gamma]} D^\mu \varphi_I(z) - D^\mu \varphi_I(x)\right||x - y|^{[\gamma]}\, dy$$

$$\leq \text{ const. } \|\varphi\|_{\mathcal{M}_\gamma} \int_{|x-y|\leq 5|x-x'|} |x - y|^{-n-[\gamma]+\gamma}|x|^{-n-2\gamma}\, dy$$

$$\leq \text{ const. } \|\varphi\|_{\mathcal{M}_\gamma}|x|^{-n-2\gamma} \int_0^{5|x-x'|} r^{-n-[\gamma]+\gamma}r^{n-1}\, dr$$

$$\leq \text{ const. } \|\varphi\|_{\mathcal{M}_\gamma}|x - x'|^{\gamma-[\gamma]}|x|^{-n-2\gamma} .$$

II is estimated in a similar way. Using lemma (4.7) again

$$III \leq \text{ const. } \|\varphi\|_{\mathcal{M}_\gamma}|x - x'|^{\delta-[\delta]}|x|^{-n-2\gamma} \int_{|x'-y|>2|x-x'|} |x' - y|^{-n-\delta}|x' - y|^\gamma\, dy$$

$$\leq \text{ const. } \|\varphi\|_{\mathcal{M}_\gamma}|x - x'|^{\delta-[\delta]}|x|^{-n-2\gamma}|x - x'|^{\gamma-\delta}$$

$$= \text{ const. } \|\varphi\|_{\mathcal{M}_\gamma}|x - x'|^{\gamma-[\gamma]}|x|^{-n-2\gamma} .$$

Finally, using lemma (4.5), we have

$$IV \leq \text{ const. } \sum_{|\alpha|+|\mu|\leq|\beta|} \left|D^{\alpha+\mu}\varphi_I(z) - D^{\alpha+\mu}\varphi_I(x')\right| |x-x'|^{|\beta|-|\alpha|}|x-x'|^{|\alpha|-|\beta|}$$

$$\leq \text{ const. } \|\varphi\|_{\mathcal{M}_\gamma}|x-x'|^{\gamma-[\gamma]}|x|^{-n-2\gamma} .$$

Now we estimate $T\varphi_J$.

$$|D^\beta(T\varphi_J)(x) - D^\beta(T\varphi_J)(x')|$$

$$= \left| \int \left[D_x^\beta K(x,y) - D_x^\beta K(x',y) \right]\varphi_J(y)\, dy \right|$$

$$\leq \left| \int \left\{ \left[D_x^\beta K(x,y) - D_x^\beta K(x',y) \right] - \sum_{|\alpha|\leq[\gamma]} \frac{D_y^\alpha D_x^\beta K(x,0) - D_y^\alpha D_x^\beta K(x',0)}{\alpha!} y^\alpha \right\} \right.$$

$$\left. \times \varphi_J(y)\, dy \right|$$

$$+ \left| \sum_{|\alpha|\leq[\gamma]} \frac{D_y^\alpha D_x^\beta K(x,0) - D_y^\alpha D_x^\beta K(x',0)}{\alpha!} \int \varphi_J(y)y^\alpha\, dy \right|$$

$$= I + II .$$

It is here that we use the double Lipschitz estimate from condition $\mathcal{H}(\delta)$.

$$I \leq \int \sum_{|\alpha|=[\gamma]} \left| \left[D_y^\alpha D_x^\beta K(x,z) - D_y^\alpha D_x^\beta K(x,0) \right] \right.$$

$$\left. - \left[D_y^\alpha D_x^\beta K(x',z) - D_y^\alpha D_x^\beta K(x',0) \right] \right| |y|^{[\gamma]}|\varphi_J(y)|\, dy$$

$$\leq \text{ const. } \|\varphi\|_{\mathcal{M}_\gamma} \int_{|y|\leq\frac{|x|}{2}} |x-x'|^{\delta-[\delta]}|y|^{\delta-[\delta]}|x|^{-n-2\delta}|y|^{[\gamma]}|y|^{-n-\gamma}\, dy$$

$$\leq \text{ const. } \|\varphi\|_{\mathcal{M}_\gamma}|x-x'|^{\delta-[\delta]}|x|^{-n-\delta-\gamma}$$

$$= \text{ const. } \|\varphi\|_{\mathcal{M}_\gamma}\left(\frac{|x-x'|}{|x|}\right)^{\delta-[\delta]}|x|^{-n-\gamma-[\gamma]}$$

$$\leq \text{ const. } \|\varphi\|_{\mathcal{M}_\gamma}\left(\frac{|x-x'|}{|x|}\right)^{\gamma-[\gamma]}|x|^{-n-\gamma-[\gamma]}$$

$$= \text{ const. } \|\varphi\|_{\mathcal{M}_\gamma}|x-x'|^{\gamma-[\gamma]}|x|^{-n-2\gamma} .$$

Using the estimate derived in Step 2 for the moments of φ_J we obtain

$$II \ \leq \ \text{const.} \sum_{|\alpha|\leq[\gamma]} |x-x'|^{\delta-[\delta]}|x|^{-n-\delta-|\alpha|} \left| \int \varphi_J(y)y^\alpha \, dy \right|$$

$$\leq \ \text{const.} |x-x'|^{\delta-[\delta]}|x|^{-n-\delta} \sum_{|\alpha|\leq[\gamma]} |x|^{-|\alpha|}\|\varphi\|_{\mathcal{M}_\gamma}|x|^{|\alpha|-\gamma}$$

$$\leq \ \text{const.} \|\varphi\|_{\mathcal{M}_\gamma}|x-x'|^{\delta-[\delta]}|x|^{-n-\delta-\gamma}$$

$$\leq \ \text{const.} \|\varphi\|_{\mathcal{M}_\gamma}|x-x'|^{\gamma-[\gamma]}|x|^{-n-2\gamma} \ .$$

Finally, we estimate

$$\left| D^\beta(T\varphi_L)(x) - D^\beta(T\varphi_L)(x') \right|$$

$$\leq \ \text{const.} \int \left| D_x^\beta K(x,y) - D_x^\beta K(x',y)\right||\varphi_L(y)| \, dy$$

$$\leq \ \text{const.} \|\varphi\|_{\mathcal{M}_\gamma} \int_{|x-y|,|x|>\frac{|x|}{2}} |x-x'|^{\delta-[\delta]}|x-y|^{-n-\delta}|y|^{-n-\gamma} \, dy$$

$$(|y| \leq |x| + |x-y| \leq 3|x-y|)$$

$$\leq \ \text{const.} \|\varphi\|_{\mathcal{M}_\gamma}|x-x'|^{\delta-[\delta]} \int_{|y|>\frac{|x|}{2}} |y|^{-n-\delta-n-\gamma} \, dy$$

$$\leq \ \text{const.} \|\varphi\|_{\mathcal{M}_\gamma}|x-x'|^{\delta-[\delta]}|x|^{-n-\gamma-\delta}$$

$$\leq \ \text{const.} \|\varphi\|_{\mathcal{M}_\gamma}|x-x'|^{\gamma-[\gamma]}|x|^{-n-2\gamma} \ .$$

This concludes the proof of theorem (4.1). $\square$

5. Triebel-Lizorkin spaces

In this section we will first motivate and then give a precise definition of the Triebel-Lizorkin scale of spaces. We will then discuss the connection between boundedness on these spaces and boundedness on molecular spaces.

Triebel-Lizorkin spaces arise from Littlewood-Paley characterizations of the classical function spaces. Recall the Littlewood-Paley *g-function*

$$g_1 f(x) \ = \ \left(\int_0^\infty \left| t\frac{\partial P_t}{\partial t} * f(x) \right|^2 \frac{dt}{t} \right)^{\frac{1}{2}}$$

where P is the Poisson kernel ([**Ste**]). It is well known that $\|g_1 f\|_p \sim \|f\|_p$ for $1 < p < \infty$. If we take $\varphi = t\frac{\partial P_t}{\partial t}\Big|_{t=1}$ we obtain

$$g_1 f(x) \ = \ \left(\int_0^\infty |\varphi_t * f(x)|^2 \frac{dt}{t} \right)^{\frac{1}{2}} \ .$$

Replacing this integral with its usual discretization $(t \sim 2^{-j})$ we obtain the approximation

$$(*) \qquad g_1 f(x) \ \sim \ \left(\sum_{j \in \mathbb{Z}} | \varphi_j * f(x) |^2 \right)^{\frac{1}{2}}$$

where $\varphi_j(x) = 2^{nj} \varphi(2^j x)$. We would then hope that the L^p-norm of the right hand side of $(*)$ is comparable to the L^p-norm of f. This motivates the following definition of Triebel-Lizorkin spaces.

Denote by $\mathcal{S}'/\mathcal{P}$ the space of tempered distributions modulo polynomial functions and let φ be a function in $\mathcal{S}(\mathbb{R}^n)$ satisfying the following conditions:

$$\operatorname{supp} \hat{\varphi} \subset \{ \xi : 1/2 \le |\xi| \le 2 \} \ , \qquad |\hat{\varphi}(\xi)| \ge c > 0 \ , \quad 3/5 \le |\xi| \le 5/3.$$

Then for each $\alpha \in \mathbb{R}$ and choice of $0 < p < \infty$, $0 < q \le \infty$ the *Triebel-Lizorkin* space $\dot{F}_p^{\alpha,q}$ consists of all elements f in $\mathcal{S}'/\mathcal{P}$ such that

$$\|f\|_{\dot{F}_p^{\alpha,q}} \ = \ \left\| \left[\sum_{j \in \mathbb{Z}} \left(2^{j\alpha} | \varphi_j * f| \right)^q \right]^{1/q} \right\|_p \ < \infty$$

This definition is independent of the choice of the function φ within equivalence of norms. Most of the classical function spaces from harmonic analysis are included in this scale of spaces. In particular, $L^p \sim \dot{F}_p^{0,2}$ for $1 < p < \infty$ and $H^p \sim \dot{F}_p^{0,2}$ for $0 < p \le 1$ (compare with $(*)$). The parameter α corresponds to smoothness of functions. For example, the Sobolev space L_α^p, $1 < p < \infty$, is equivalent to $\dot{F}_p^{\alpha,2}$.

To define $\dot{F}_p^{\alpha,q}$ for $p = \infty$ we do not simply replace the L^p-norm with an L^∞-norm. There are several problems with this naive approach. In particular, the space is no longer independent of the function φ. Moreover, the expected equivalence $\dot{F}_\infty^{0,2} \sim BMO$ does not hold. For the correct definition, we define $\dot{F}_\infty^{\alpha,q}$ to be the space of all f in $\mathcal{S}'/\mathcal{P}$ such that

$$\|f\|_{\dot{F}_\infty^{\alpha,q}} \ = \ \sup_P \left(\frac{1}{|P|} \int_P \sum_{j=-\log_2 |P|^{\frac{1}{n}}}^{\infty} \left(2^{j\alpha} | \varphi_j * f(x)| \right)^q dx \right)^{\frac{1}{q}} \ < \infty$$

where the supremum is taken over all dyadic cubes P. With this definition we have $\dot{F}_\infty^{0,2} \sim BMO$.

There are companion families of sequence spaces $\dot{f}_p^{\alpha,q}$ which play the same role for $\dot{F}_p^{\alpha,q}$ as ℓ^2 does for $L^2(\mathbb{R}^n)$. Let $\dot{f}_p^{\alpha,q}$ $(\alpha \in \mathbb{R}, 0 < p < \infty, 0 < q \le \infty)$ be the collection of all sequences $s = \{s_Q\}_Q$ over dyadic cubes such that

$$\|s\|_{\dot{f}_p^{\alpha,q}} \ = \ \left\| \left(\sum_Q (|Q|^{-\frac{\alpha}{n}} |s_Q| \tilde{\chi}_Q)^q \right)^{\frac{1}{q}} \right\|_p \ < \infty$$

where $\tilde{\chi}_Q = |Q|^{-\frac{1}{2}}\chi_Q$ is the L^2-normalized characteristic function of Q. The corresponding sequence space for $p = \infty$ is defined in terms of the norm

$$\|s\|_{\dot{f}^{\alpha,q}_\infty} = \sup_P \left(\frac{1}{|P|} \int_P \sum_{Q \subset P} (|Q|^{-\frac{\alpha}{n}} |s_Q| \tilde{\chi}_Q(x))^q \, dx \right)^{\frac{1}{q}}$$

where the supremum and sum are over dyadic cubes.

We will now show that any operator mapping $\mathcal{M}_\delta$-molecules to $\mathcal{M}_\delta$-molecules is automatically bounded on certain Triebel-Lizorkin spaces.

(5.1) THEOREM. *Let T be a linear operator that is bounded from $\mathcal{M}^{(0)}_\delta(\mathbb{R}^n)$ into itself independently of center and width. Then T extends to a bounded linear operator on $\dot{F}^{\alpha,q}_p$, $0 < p, q \leq \infty$, $\alpha > 0$, whenever*

$$\delta > n\left(\frac{1}{\min(1,p,q)} - 1\right), \quad \delta > n\left(\frac{1}{\min(1,p,q)} - 1\right) - \alpha, \quad \delta > \alpha \ .$$

In particular we obtain the following results:

(5.2) COROLLARY. *Suppose $1 \leq p, q \leq \infty$ and $|\alpha| < \delta$, then T has a continuous extension to $\dot{F}^{\alpha,q}_p$.*

(5.3) COROLLARY. *Suppose $\alpha \geq 0$ and $0 < p \leq \min(1,q)$. If*

$$\delta > n\left(\frac{1}{p} - 1\right), \qquad \delta > \alpha$$

and T is bounded on $\mathcal{M}^{(0)}_\delta(\mathbb{R}^n)$, then T has a continuous extension to $\dot{F}^{\alpha,q}_p$.

The proof of this theorem relies on the Frazier-Jawerth molecular decomposition of $\dot{F}^{\alpha,q}_p$ ([**Tor**]). For $\alpha \in \mathbb{R}$ and $0 < p, q \leq \infty$, set

$$J = \frac{n}{\min(1,p,q)}, \quad L = \max([J - n - \alpha], [J - n]) \ .$$

A smooth atom for $\dot{F}^{\alpha,q}_p$ associated with the unit cube Q_0 is a function $a \in \mathcal{D}(\mathbb{R}^n)$ satisfying

$$\text{supp } a \subset 3Q_0, \qquad \int_{\mathbb{R}^n} x^\gamma a(x) \, dx = 0 \quad (|\gamma| \leq L),$$

$$|D^\gamma a(x)| \leq 1 \ , \quad (\, |\gamma| \leq \max([\alpha], 0) + 1 \,) \ .$$

Let $M > J$ and $1 > \eta > \alpha - [\alpha]$. A smooth (M, η)-molecule for $\dot{F}_p^{\alpha, q}$ is a function m satisfying

$$|m(x)| \leq \frac{1}{(1 + |x|)^{\max(M, M-\alpha)}},$$

$$\int x^\gamma m(x)\, dx = 0, \quad (|\gamma| \leq [J - n - \alpha]),$$

$$|D^\gamma m(x)| \leq \frac{1}{(1 + |x|)^M}, \quad (|\gamma| \leq [\alpha]),$$

$$|D^\gamma m(x) - D^\gamma m(x')| \leq |x - x'|^\eta \sup_{|z| \leq |x-x'|} \frac{1}{(1 + |z - x|)^M}.$$

It is easy to verify that, for δ satisfying the inequalities in the theorem, we have

$$\text{smooth atoms} \subseteq \mathcal{M}_\delta\text{-molecules with norm } 1 \subseteq \text{smooth molecules.}$$

where $M = n + \delta$ and $\eta = \delta - [\delta]$. Frazier and Jawerth show that any operator that maps smooth atoms to smooth molecules (with an appropriate normalization constant) has a bounded extension to the appropriate Triebel-Lizorkin space. The result now follows.

CHAPTER 3

Frames

1. Continuity of group action

For a function $\psi \in \mathcal{M}_\delta(\mathbb{R}^n)$ it is not necessarily true that

$$\lim_{w \to e} \|(I - U(w))\psi\|_{\mathcal{M}_\delta} = 0$$

for the same reason that a bounded, continuous function on $\mathbb{R}^n$ need not be uniformly continuous (see section 2 of chapter 1 for definitions). But the limit will be zero if it is taken in $\mathcal{M}_\gamma(\mathbb{R}^n)$ for $\gamma < \delta$.

(1.1) THEOREM. *For each fixed φ in $\mathcal{M}_\delta(\mathbb{R}^n)$ and each γ, $0 < \gamma < \delta$,*

$$\lim_{w \to e} \|(I - U(w))\varphi\|_{\mathcal{M}_\gamma} = 0.$$

PROOF OF THEOREM (1.1). Suppose $w = (r, s)$, then for any multi-index α with $|\alpha| < \delta$ we have

$$D^\alpha\{(I - U(w))\varphi\}(x) = D^\alpha\varphi(x) - r^{|\alpha|}U(w)D^\alpha\varphi(x)$$

$$= (D^\alpha\varphi(x) - D^\alpha\varphi(rx + s)) + (1 - r^{\frac{n}{2}+|\alpha|})D^\alpha\varphi(rx + s) .$$

Assume for simplicity that $|\alpha| = [\delta] < \gamma < \delta$. The other cases are handled similarly. Since $\varphi \in \mathcal{M}_\delta$ we have

$$|D^\alpha\varphi(x) - D^\alpha\varphi(x')| \leq \text{const.} \|\varphi\|_{\mathcal{M}}|x - x'|^{\delta-[\delta]}\left\{\frac{1}{(1 + |x|)^{n+2\delta}} + \frac{1}{(1 + |x'|)^{n+2\delta}}\right\}$$

for all $x, x' \in \mathbb{R}^n$ (see lemma (2.16) in chapter 1). Thus

$$|D^\alpha\{(I - U(w))\varphi\}(x)|$$

$$\leq |D^\alpha\varphi(x) - D^\alpha\varphi(rx + s)| + (r^{\frac{n}{2}+|\alpha|} - 1)|D^\alpha\varphi(rx + s)|$$

$$\leq \text{const.} \|\varphi\|_{\mathcal{M}} |(r - 1)x + s|^{\delta-[\delta]}\left\{\frac{1}{(1 + |x|)^{n+2\delta}} + \frac{1}{(1 + |rx + s|)^{n+2\delta}}\right\}$$

$$+ (r^{\frac{n}{2}+|\alpha|} - 1)\frac{\|\varphi\|_{\mathcal{M}}}{(1 + |rx + s|)^{n+\delta+[\delta]}}.$$

Given $\varepsilon_0 > 0$ therefore, there exists $r_0 > 1$ and $s_0 > 0$ so that

$$(1.2) \quad |D^\alpha\{(I - U(w))\varphi\}(x)| \leq \varepsilon_0\|\varphi\|_{\mathcal{M}_\delta}(1 + |x|)^{-n-\delta-[\delta]} \quad \begin{matrix} 1 < r < r_0, \\ 0 < s_\ell < s_0, \end{matrix}$$

43

Now we treat the Lipschitz condition. Using similar estimates as before it is not hard to show that

$$|D^\alpha\{(I-U(w))\varphi\}(x) - D^\alpha\{(I - U(w))\varphi\}(x')|$$

$$\leq |D^\alpha\varphi(x) - D^\alpha\varphi(x')| + |D^\alpha U(w)\varphi(x) - D^\alpha U(w)\varphi(x')|$$

$$\leq \text{const.}\,|x - x'|^{\delta-[\delta]}(1 + |x|)^{-n-2\delta}$$

for $|x - x'| \leq \frac{1}{2}(1+|x|)$ and w in some compact neighborhood of the identity. The constant does not necessarily tend to zero as $w \to e$. However, using (1.2) we obtain

$$|D^\alpha\{(I - U(w))\varphi\}(x) - D^\alpha\{(I - U(w))\varphi\}(x')|$$

$$\leq |D^\alpha\{(I - U(w))\varphi\}(x)| + |D^\alpha\{(I - U(w))\varphi\}(x')|$$

$$\leq \text{const.}\,\varepsilon_0\|\varphi\|_{\mathcal{M}_\delta}(1 + |x|)^{-n-\delta-[\delta]} + \text{const.}\,\varepsilon_0\|\varphi\|_{\mathcal{M}_\delta}(1 + |x'|)^{-n-\delta-[\delta]}$$

$$\leq \text{const.}\,\varepsilon_0\|\varphi\|_{\mathcal{M}_\delta}(1 + |x|)^{-n-\delta-[\delta]}$$

for $|x - x'| \leq \frac{1}{2}(1+|x|)$. Now, taking a geometric mean $(A = A^\eta A^{1-\eta}, 0 \leq \eta \leq 1)$ of the previous two estimates we obtain

$$|D^\alpha\{(I - U(w))\varphi\}(x) - D^\alpha\{(I - U(w))\varphi\}(x')|$$

$$\leq \text{const.}\,\varepsilon_0|x - x'|^{\gamma-[\gamma]}(1 + |x|)^{-n-\delta-\gamma}$$

for $|x - x'| \leq \frac{1}{2}(1+|x|)$. Therefore we have

$$\lim_{w \to e} \|(I - U(w))\varphi\|_{\mathcal{M}_\gamma} = 0$$

completing the proof. $\qquad\square$

2. Inversion of frame operator

We now show that for sufficiently small mesh size the frame operator is invertible on certain molecule spaces. This will imply its invertibility on many other spaces including $L^p(\mathbb{R}^n), H^1(\mathbb{R}^n)$ and many Triebel-Lizorkin spaces. Since ψ is admissible (this will be relaxed in section 4 of this chapter),

$$f - |\Delta|\mathcal{D}_{\psi,\psi}f = \int_H \langle f, U^*(z)\psi\rangle U^*(z)\psi\,dz$$

$$- \sum_{j,k}\left(\int_H \chi_\Delta(z\lambda_{j,k}^{-1})\,dz\right)\langle f, U^*(\lambda_{j,k})\psi\rangle U^*(\lambda_{j,k})\psi\,.$$

In turn, the right hand side can be written as a sum

$$\mathcal{T}_{a\beta}f + \mathcal{T}_{\alpha b}f = \sum_{j,k}\int_H \chi_\Delta(z\lambda_{j,k}^{-1})\langle f, U^*(z)\psi\rangle\{U^*(z) - U^*(\lambda_{j,k})\}\psi\,dz$$

$$+ \sum_{j,k}\int_H \chi_\Delta(z\lambda_{j,k}^{-1})\langle f, \{U^*(z) - U^*(\lambda_{j,k})\}\psi\rangle U^*(\lambda_{j,k})\psi\,dz$$

of two operators $\mathcal{T}_{a\beta}$ and $\mathcal{T}_{\alpha b}$ of Cotlar type. To define the first of these set

$$a(z\,;\,y) = \psi(y), \qquad \beta(z\,;\,x) = \sum_{j,k}\chi_\Delta(z\lambda_{j,k}^{-1})\{I - U(z\lambda_{j,k}^{-1})\}\psi(x)\ .$$

Clearly $a, \beta : H \to \mathcal{M}_\delta(\mathbb{R}^n)$ are L^∞-bounded functions having vanishing moment since ψ is an $\mathcal{M}_\delta$-molecule. Moreover, by corollary (3.6) we have

$$\|\mathcal{T}_{a\beta}f\|_{\mathcal{M}_\gamma} \leq \text{const.}\, \|a\|_{\infty,\gamma}\|\beta\|_{\infty,\gamma}\|f\|_{\mathcal{M}_\gamma}$$

$$\leq \text{const.}\, \|\psi\|_{\mathcal{M}_\gamma} \sup_{w\in\Delta} \|(I - U(w))\psi\|_{\mathcal{M}_\gamma}\|f\|_{\mathcal{M}_\gamma}$$

for each γ, $\gamma < \delta$. The continuity of the representation $w \to U(w)$ now plays a crucial role. Fix $\gamma < \delta$ and $\varepsilon > 0$. Then by theorem (1.1),

$$(2.1) \qquad \|\mathcal{T}_{a\beta}f\|_{\mathcal{M}_\gamma} \leq \text{const.}\,\varepsilon\left(\|\psi\|_{\mathcal{M}_\gamma}\right)^2\|f\|_{\mathcal{M}_\gamma}$$

provided the mesh size (r, s) defining is restricted to $1 < r < r_0$ and $0 < s_\ell < s_0$.

To define the second of the operators, set

$$\alpha(z\,;\,y) = \sum_{j,k}\chi_\Delta(z\lambda_{j,k}^{-1})\{I - U(z\lambda_{j,k}^{-1})\}\psi(y)$$

and

$$b(z\,;\,x) = \sum_{j,k}\chi_\Delta(z\lambda_{j,k}^{-1})U(z\lambda_{j,k}^{-1})\psi(x).$$

As before these are L^∞-bounded functions having vanishing moment. Consequently, $\mathcal{T}_{\alpha b}$ is a Cotlar operator and

$$\|\mathcal{T}_{\alpha b}f\|_{\mathcal{M}_\gamma} \leq \text{const.}\, \|\alpha\|_{\infty,\gamma}\|b\|_{\infty,\gamma}\|f\|_{\mathcal{M}_\gamma}$$

$$\leq \text{const.}\, \sup_{w\in\Delta} \|U(w)\psi\|_{\mathcal{M}_\gamma} \sup_{w\in\Delta} \|(I - U(w))\psi\|_{\mathcal{M}_\gamma}\ .$$

Since $w \to U(w)$ is uniformly bounded on compact subsets of H,

$$\sup_{w\in\Delta} \|U(w)\psi\|_{\mathcal{M}_\gamma} \leq 2\,\|\psi\|_{\mathcal{M}_\gamma}$$

so long as $1 < r < r_0$ and $0 < s_j < s_0$. Together with the continuity of $w \to U(w)$ this ensures that

$$\|\mathcal{T}_{\alpha b}f\|_{\mathcal{M}_\gamma} \leq \text{const.}\,\varepsilon\left(\|\psi\|_{\mathcal{M}_\gamma}\right)^2\|f\|_{\mathcal{M}_\gamma}$$

whenever $1 < r < r_0$ and $0 < s_\ell < s_0$, choosing possibly smaller values of (r_0, s_0). Combining this with estimate (2.1) we see that

$$\|f - |\Delta|\mathcal{D}_{\psi,\psi}f\|_{\mathcal{M}_\gamma} \leq \text{const.}\,\varepsilon\left(\|\psi\|_{\mathcal{M}_\gamma}\right)^2\|f\|_{\mathcal{M}_\gamma}$$

provided $1 < r < r_0$ and $0 < s_j < s_0$ (note that $|\Delta| = s_0^n \ln r_0$). Thus the Neumann series

$$\mathcal{D}_{\psi,\psi}^{-1} = |\Delta|\sum_{n=0}^{\infty}(I - |\Delta|\mathcal{D}_{\psi,\psi})^n$$

will converge for all sufficiently small mesh size proving that the frame operator is invertible on $\mathcal{M}_\gamma$.

We can now prove theorem (1.1.5). Consider the frame operator

$$\mathcal{D}f \;=\; \mathcal{D}_{\psi,\psi}f \;=\; \sum_{j\in\mathbb{Z}}\sum_{k\in\mathbb{Z}^n} \langle f,\psi_{j,k}\rangle\psi_{j,k}.$$

We now apply $\mathcal{D}^{-1}$ (for suitably small mesh size) to both sides of this equation. We will see in the next section that this sum converges in $\dot{F}_p^{\alpha,q}$ so we can pass $\mathcal{D}^{-1}$ through the sums and obtain

$$f \;=\; \sum_{j\in\mathbb{Z}}\sum_{k\in\mathbb{Z}^n} \langle f,\psi_{j,k}\rangle\rho_{j,k}$$

where $\rho_{j,k} \equiv \mathcal{D}^{-1}\psi_{j,k} = \mathcal{D}^{-1}U^*(\lambda_{jk})\psi$. Since $\mathcal{D}^{-1}$ does not in general commute with $U^*(w)$ we can not assert that $\rho_{j,k} = U^*(\lambda_{jk})\rho$ for some function ρ.

The operator $\mathcal{D}^{-1}$ is bounded on $\mathcal{M}_\gamma$ so the function $\rho_{j,k}$ belongs to an $\mathcal{M}_\gamma$ space with the expected width and center. Thus the results of the next section show that this series converges in the appropriate Triebel-Lizorkin spaces.

3. Convergence in Triebel-Lizorkin spaces

In this section we will see that the sum defining a frame operator associated to a function φ converges in certain Triebel-Lizorkin spaces depending on the molecule space that φ belongs to. To do this we will need to some results of Frazier and Jawerth ([**FJ**]).

Fix $\alpha\in\mathbb{R}$ and $0<p,q\le\infty$. Let $J = n/\min(1,p,q)$. Choose $M>J$ and $\alpha-[\alpha]<\eta\le 1$. A function m is a Type A (M,η)-molecule for $\dot{F}_p^{\alpha,q}$ if

$$|m(x)| \le (1+|x|)^{-\max(M,M-\alpha)}$$

$$\int x^\gamma m(x)\,dx = 0 \quad |\gamma|\le[J-n-\alpha]$$

$$|D^\gamma m(x)| \le (1+|x|)^{-M} \quad |\gamma|\le[\alpha]$$

$$|D^\gamma m(x)-D^\gamma m(y)| \le |x-y|^\eta \sup_{|z|\le|x-y|}(1+|z-x|)^{-M} \quad |\gamma|=[\alpha]\,.$$

Now suppose ρ is chosen such that $(J-\alpha)-[J-\alpha]<\rho\le 1$. A function b is a Type B (M,ρ)-molecule for $\dot{F}_p^{\alpha,q}$ if

$$|b(x)| \le (1+|x|)^{-\max(M,M+n+\alpha-J)}$$

$$\int x^\gamma b(x)\,dx = 0 \quad |\gamma|\le[\alpha]$$

$$|D^\gamma b(x)| \le (1+|x|)^{-M} \quad |\gamma|\le[J-n-\alpha]$$

$$|D^\gamma b(x)-D^\gamma b(y)| \le |x-y|^\rho \sup_{|z|\le|x-y|}(1+|z-x|)^{-M} \quad |\gamma|=[J-n-\alpha]\,.$$

(Notice the inversion of the roles of $J-n-\alpha$ and α as compared to Type A molecules). We define a molecule m_Q associated to a dyadic cube Q by dilating and translating appropriately.

The following two results are due to Frazier and Jawerth ([**FJ**]).

(3.1) **THEOREM (FRAZIER-JAWERTH).** *If $f = \sum_Q s_Q m_Q$, where $\{s_Q\} \in \dot{f}_p^{\alpha,q}$ and $\{m_Q\}$ is a family of Type A molecules for $\dot{F}_p^{\alpha,q}$, then*

$$\|f\|_{\dot{F}_p^{\alpha,q}} \leq \text{const.} \|\{s_Q\}\|_{\dot{f}_p^{\alpha,q}} \ .$$

(3.2) **THEOREM (FRAZIER-JAWERTH).** *If $f \in \dot{F}_p^{\alpha,q}$ and $\{b_Q\}$ is a family of Type B molecules for $\dot{F}_p^{\alpha,q}$, then*

$$\|\{\langle f, b_Q \rangle\}\|_{\dot{f}_p^{\alpha,q}} \leq \text{const.} \|f\|_{\dot{F}_p^{\alpha,q}} \ .$$

The molecule spaces $\mathcal{M}_\delta^{(0)}(\mathbb{R}^n)$ include both the Type A and Type B molecules simultaneously.

(3.3) **THEOREM.** *Suppose $0 < p, q \leq \infty$, $\alpha \in \mathbb{R}$ and δ are chosen so that*

$$\delta > n \left(\frac{1}{\min(1,p,q)} - 1 \right), \quad \delta > n \left(\frac{1}{\min(1,p,q)} - 1 \right) - \alpha, \quad \delta > \alpha \ .$$

Then an $\mathcal{M}_\delta$-molecule φ with $\|\varphi\|_{\mathcal{M}_\delta} \leq 1$ is both a Type A and a Type B molecule for $\dot{F}_p^{\alpha,q}$.

PROOF. This is mostly self-explanatory. We will prove the size estimate for Type B molecules. Choose $M = \min(J + \delta - \alpha, n + \delta)$. Then it is easy to check that $M > J$. It is also easy to see that $n + \delta \geq \max(M, M + n + \alpha - J)$. This implies, since $\|\varphi\|_{\mathcal{M}_\delta} \leq 1$, that

$$|\varphi(x)| \leq (1 + |x|)^{-n-\delta} \leq (1 + |x|)^{-\max(M, M+n+\alpha-J)}$$

The other conditions are similar. $\qquad\qquad\qquad\qquad\qquad\qquad\qquad\qquad\square$

We can now state the convergence result. This theorem generalizes to the case where the analyzing and synthesizing functions are not necessarily the same. The functions need not even be translations and dilations of a single function, but just a family of functions that satisfy the appropriate molecule estimates.

(3.4) **COROLLARY.** *Suppose $0 < p, q \leq \infty$, $\alpha \in \mathbb{R}$ and δ are chosen such that*

$$\delta > n \left(\frac{1}{\min(1,p,q)} - 1 \right), \quad \delta > n \left(\frac{1}{\min(1,p,q)} - 1 \right) - \alpha, \quad \delta > \alpha \ .$$

Then for any $\varphi \in \mathcal{M}_\delta^{(0)}(\mathbb{R}^n)$ the series

$$\sum_Q \langle f, \varphi_Q \rangle \varphi_Q \ ,$$

converges in $\dot{F}_p^{\alpha,q}$ for any $f \in \dot{F}_p^{\alpha,q}$.

4. Algebras of singular integral operators

In this section we will show that a certain class of singular integral operators form an algebra. Moreover, we will show that many of its elements are invertible in this algebra. This will enable us to prove our results using the more general admissibility condition introduced in chapter 1.

Suppose $\psi \in \mathcal{M}_\delta^{(0)}$ satisfies

$$(*) \qquad A \leq \int_0^\infty |\hat{\psi}(t\xi)|^2 \, \frac{dt}{t} \leq B, \qquad \forall \xi \in S^{n-1}$$

where $A, B > 0$. Define

$$(**) \qquad Tf(x) = \int_0^\infty \psi_t * \tilde{\psi}_t * f(x) \, \frac{dt}{t}.$$

Then, by Plancherel's theorem, T is invertible on $L^2(\mathbb{R}^n)$. Moreover, we have the following result.

(4.1) THEOREM. *For any $\eta < \delta$, T^{-1} is bounded from $\mathcal{M}_\eta^{(0)}$ into itself.*

In particular, we obtain a more general version of the Calderón reproducing formula.

(4.2) COROLLARY. *Suppose $\psi \in \mathcal{M}_\gamma^{(0)}$ satisfies $(*)$, then for any $\eta < \delta$ there exists a function $\rho \in \mathcal{M}_\eta^{(0)}$ such that*

$$f = \int_0^\infty \rho_t * \tilde{\psi}_t * f \, \frac{dt}{t}$$

for all $f \in L^2$.

PROOF. This follows from applying T^{-1} to both sides of $(**)$ (note that T^{-1} is translation and dilation invariant). $\qquad\qquad\square$

Earlier in this chapter, while proving theorem (1.1.5), we used the admissibility of the function ψ ($A = B = 1$ in $(*)$). The proof can now be extended to the weaker admissibility condition by using the reproducing formula from the preceding corollary.

We now start the proof of theorem (4.1). Assume without loss of generality that $[\delta] < \eta < \delta$. We will first show that the kernel of T lies in a certain Lipschitz space. Choose $\ell \in \mathbb{N}$ and $0 < \varepsilon < 1$ such that $\eta < \ell + \varepsilon < \delta$. From theorem (3.1) of chapter 2 we know that $T \in CZO(\ell + \varepsilon)$ and the kernel

$$K(x) = \int_0^\infty \psi_t * \tilde{\psi}_t(x) \, \frac{dt}{t}$$

satisfies a double-Lipschitz condition of order $\ell + \varepsilon$. Therefore

$$|D_x^\gamma D_y^\mu K(x - y)| \leq \text{const.} \, |x - y|^{-n-|\gamma|-|\mu|}$$

for $|\gamma|, |\mu| \leq \ell$. Thus

$$|D^\gamma K(x)| \leq \text{const.} \, |x|^{-n-|\gamma|}$$

for $|\gamma| \leq 2\ell$ and hence $|D^\gamma K(x)| \leq C$ for $x \in S^{n-1}$ and $|\gamma| \leq 2\ell$. From the double-Lipschitz condition we obtain

$$|D^\gamma K(x + t) + D^\gamma K(x - t) - 2D^\gamma K(x)| \leq \text{const.} \, |t|^{2\varepsilon} |x|^{-n-2\ell-2\varepsilon}$$

for $|t| \le \frac{1}{3}|x|$ and $|\gamma| = 2\ell$. Thus

$$|D^\gamma K(x+t) + D^\gamma K(x-t) - 2D^\gamma K(x)| \;\le\; \text{const.}\,|t|^{2\varepsilon}$$

for $x \in S^{n-1}$ and $|\gamma| = 2\ell$.

We would like to conclude from this that $K \in \Lambda_{2(\ell+\varepsilon)}(S^{n-1})$, the Lipschitz space of order $2(\ell + \varepsilon)$ on S^{n-1}. To do this we use a result from [**Ste**], pg. 146.

(4.3) PROPOSITION. *Suppose $0 < \alpha < 2$. Then $f \in \Lambda_\alpha(\mathbb{R}^n)$ if and only if $f \in L^\infty(\mathbb{R}^n)$ and*

$$\| f(x + t) + f(x - t) - 2f(x) \|_\infty \;\le\; \text{const.}\,|t|^\alpha \; .$$

From this proposition (note that $0 < 2\varepsilon < 2$) we see that $D^\gamma K \in \Lambda_{2\varepsilon}(S^{n-1})$ for $|\gamma| = 2\ell$ (the changes needed to replace $\mathbb{R}^n$ with S^{n-1} are routine). Thus we can conclude that $K \in \Lambda_{2(\ell+\varepsilon)}(S^{n-1})$.

Now choose ε' and ε'' so that $\eta < \ell + \varepsilon'' < \ell + \varepsilon' < \ell + \varepsilon < \delta$. Then there exists $p < \infty$ such that $2(\ell + \varepsilon') - \frac{n-1}{p} > 2(\ell + \varepsilon'')$. Since any power of the Laplacian $(-\Delta)^\alpha$ maps Λ_β into $\Lambda_{\beta-\alpha}$ for $\beta > \alpha$ and we have $L^r(S^{n-1}) \hookrightarrow L^s(S^{n-1})$ for $r \ge s$ we get the embedding $\Lambda_{2(\ell+\varepsilon)}(S^{n-1}) \hookrightarrow L^p_{2(\ell+\varepsilon')}(S^{n-1})$. Therefore K belongs to the Sobolev space $L^p_{2(\ell+\varepsilon')}(S^{n-1})$.

(4.4) DEFINITION. Define $\mathcal{A}$ to be the collection of all distributions L of the form $a\delta + L'$ where δ is the Dirac mass and L' is a function homogeneous of degree $-n$, and $L' \in L^p_{2(\ell+\varepsilon')}(S^{n-1})$ with mean zero on S^{n-1}.

The space $\mathcal{A}$ is a Banach algebra with norm $\|L\|_{\mathcal{A}} = |a| + \|L'\|_{L^p_{2(\ell+\varepsilon')}(S^{n-1})}$ (see [**Cal**]). Standard Banach algebra techniques now show that the kernel K^- of the inverse of T belongs to $\mathcal{A}$. This is done by identifying all algebra homomorphisms with point evaluations of $\hat{K}$ on the unit sphere. Since $\hat{K}$ does not vanish on the unit sphere we see that $\rho(K) \ne 0$ for any homomorphism ρ. A classical theorem from the theory of Banach algebras ([**Kat**]) implies that K is invertible within the algebra $\mathcal{A}$. Thus $K^- \in L^p_{2(\ell+\varepsilon')}(S^{n-1})$.

From our choice of p it follows from the Sobolev embedding theorem that $K^- \in \Lambda_{2(\ell+\varepsilon'')}(S^{n-1})$. In particular, we see that K^- is a Calderón-Zygmund kernel of order $\ell + \varepsilon''$. Since T^{-1}, being translation invariant, annihilates polynomials up to order ℓ (proposition 2.2.17 in [**Tor**]), the boundedness of T^{-1} on $\mathcal{M}^{(0)}_\eta$ will follow once we show that K^- satisfies a double-Lipschitz condition of order $\ell + \varepsilon''$. We know that

$$|D^\gamma K^-(x+t) + D^\gamma K^-(x-t) - 2D^\gamma K^-(x)| \;\le\; \text{const.}\,|t|^{2\varepsilon''}$$

for $|\gamma| = 2\ell$ since $D^\gamma K^- \in \Lambda_{2\varepsilon''}$. We need K^- to satisfy a similar inequality that is asymmetric. The following lemma is adapted from Stein's proof of proposition (4.3) ([**Ste**]). For simplicity we do this in the $\mathbb{R}^n$ setting.

(4.5) LEMMA. *Suppose $0 < \alpha < 2$. Then $f \in \Lambda_\alpha$ if and only if $f \in L^\infty(\mathbb{R}^n)$ and*

$$|f(x + h) - f(x) + f(x - l) - f(x + h - l)| \;\le\; \text{const.}\,|h|^{\frac{\alpha}{2}}|l|^{\frac{\alpha}{2}} \; .$$

PROOF. Note that if $h = l$ this is just proposition (4.3). Thus one direction is already proved. For the converse, write $(\Delta_{h,l}F)(x) = F(x+h) - F(x) + F(x-l) - F(x+h-l)$, and observe that if F has two continuous derivatives, then

$$\Delta_{h,l}F(x) = \int_0^{|h|} \int_0^{|l|} \frac{d^2}{d\eta\, d\tau} F(x - l'\eta + h'\tau)\, d\eta\, d\tau$$

where $h' = h/|h|$ and $l' = l/|l|$. It follows immediately that

$$\|\Delta_{h,l}F\|_\infty \leq |h|\,|l| \sum_{i,j} \left\| \frac{\partial^2 F}{\partial x_i\, \partial x_j} \right\|_\infty .$$

As in Stein's treatment we have

$$f(x) = u(x,0) = \int_0^y y' \frac{\partial^2}{(\partial y')^2} u(x,y')\, dy' - y\frac{\partial u}{\partial y}(x,y) + u(x,y)$$

for all $y > 0$ where u is the Poisson extension of f to the upper-half space. Since $f \in \Lambda_\alpha$ we have the inequalities

$$\left\| \frac{\partial^2 u}{\partial x_i\, \partial x_j} \right\|_\infty \leq \text{const.}\, y^{-2+\alpha}, \qquad \left\| \frac{\partial^3 u}{\partial y\, \partial x_i\, \partial x_j} \right\|_\infty \leq \text{const.}\, y^{-3+\alpha} .$$

Thus we have

$$\|\Delta_{h,l}f\|_\infty \leq 4 \left\| \int_0^y y' \frac{\partial^2}{(\partial y')^2} u(x,y')\, dy' \right\|_\infty$$

$$+ \text{const.}\, |h|\,|l|\, y\, y^{-3+\alpha} + \text{const.}\, |h|\,|l|\, y^{-2+\alpha}$$

$$\leq \text{const.} \left\{ \int_0^y y' (y')^{-2+\alpha}\, dy' + |h|\,|l|\, y^{-2+\alpha} \right\}$$

$$\leq \text{const.} \left\{ y^\alpha + |h|\,|l|\, y^{-2+\alpha} \right\}$$

Taking $y = \sqrt{|h|\,|l|}$ we obtain the desired inequality. $\square$

To conclude the proof of theorem (4.1) we need the following proposition.

(4.6) PROPOSITION. *Suppose $T : f \to p.v.K * f$ where K is homogeneous of degree $-n$. Then $T \in CZO(\delta) \cap \mathcal{H}(\delta)$ if and only if $K \in \Lambda_{2\delta}(S^{n-1})$.*

PROOF. If $T \in CZO(\delta) \cap \mathcal{H}(\delta)$ then it follows immediately from kernel estimates and lemma (4.5) that $K \in \Lambda_{2\delta}(S^{n-1})$.

For the other direction we assume, for simplicity, that $0 < \delta < 1$. The size condition is trivial:

$$|K(x)| = |x|^{-n} \left| K\left(\frac{x}{|x|} \right) \right| \leq |x|^{-n} \|K\|_{L^\infty(S^{n-1})} .$$

We now want to show that

$$|K(x) - K(x')| \leq \text{const.}\, |x - x'|^\varepsilon |x|^{-n-\delta}$$

for $|x - x'| \leq \frac{1}{2}|x|$. We first remark that $|x| \sim |x'|$. Now

$$|K(x) - K(x')| \leq \left|K(x) - K\left(\frac{|x|}{|x'|}x'\right)\right| + \left|K\left(\frac{|x|}{|x'|}x'\right) - K(x')\right|$$

$$\leq |x|^{-n}\left|K\left(\frac{x}{|x|}\right) - K\left(\frac{x'}{|x'|}\right)\right| + \left||x|^{-n} - |x'|^{-n}\right| \|K\|_{L^\infty(S^{n-1})}$$

$$\leq |x|^{-n}\left|\frac{x}{|x|} - \frac{x'}{|x'|}\right|^\delta \|K\|_{\Lambda_\delta(S^{n-1})} + \text{const.} |x - x'||x|^{-n-1}\|K\|_{L^\infty(S^{n-1})}$$

$$\leq \text{const.} |x - x'|^\delta |x|^{-n-\delta}$$

since

$$\left|\frac{x}{|x|} - \frac{x'}{|x'|}\right| \leq \frac{|x - x'|}{|x|} + \frac{|x'|\,||x| - |x'||}{|x|\,|x'|} \ .$$

Finally, we need to show that

$$|K(x + h) - K(x) + K(x - l) - K(x + h - l)| \leq \text{const.} |h|^\delta |l|^\delta |x|^{-n-2\delta}$$

for $|h|, |l| \leq \frac{1}{3}|x|$. The simple argument used to treat the Lipschitz condition will not work for this double-Lipschitz condition. We first write

$$|K(x+h) - K(x) + K(x - l) - K(x + h - l)|$$

$$= |x|^{-n}\left|K\left(\frac{x + h}{|x|}\right) - K\left(\frac{x}{|x|}\right) + K\left(\frac{x - l}{|x|}\right) - K\left(\frac{x + h - l}{|x|}\right)\right|$$

and note that each of the four arguments lie in the annulus $A = \{\frac{1}{3} \leq |x| \leq \frac{5}{3}\}$ since $|h|, |l| \leq \frac{1}{3}|x|$. Define $\tilde{K}$ to be the homogeneous of degree 0 extension of K from S^{n-1} to A. Then $\tilde{K} \in \Lambda_{2\delta}(A)$. Define $W(x) = |x|^{-n}$, then W is C^∞ on A with each derivative belonging to $L^\infty(A)$. Thus $W \cdot \tilde{K} \in \Lambda_{2\delta}(A)$ and hence, by lemma (4.4), $W \cdot \tilde{K}$ satisfies a double-Lipschitz condition. Therefore,

$$|K(x+h) - K(x) + K(x - l) - K(x + h - l)|$$

$$= |x|^{-n}\left|K\left(\frac{x + h}{|x|}\right) - K\left(\frac{x}{|x|}\right) + K\left(\frac{x - l}{|x|}\right) - K\left(\frac{x + h - l}{|x|}\right)\right|$$

$$= |x|^{-n}\left|W\left(\frac{x + h}{|x|}\right)\tilde{K}\left(\frac{x + h}{|x|}\right) - W\left(\frac{x}{|x|}\right)\tilde{K}\left(\frac{x}{|x|}\right)\right.$$

$$\left. + W\left(\frac{x - l}{|x|}\right)\tilde{K}\left(\frac{x - l}{|x|}\right) - W\left(\frac{x + h - l}{|x|}\right)\tilde{K}\left(\frac{x + h - l}{|x|}\right)\right|$$

$$\leq C|x|^{-n}\left|\frac{h}{|x|}\right|^\delta \left|\frac{l}{|x|}\right|^\delta = C|h|^\delta |l|^\delta |x|^{-n-2\delta} \ .$$

$$\square$$

Since $K^- \in \Lambda_{2\delta}(S^{n-1})$ and $\eta < \delta$ we can apply theorem (4.1) of chapter 2 to get the boundedness of $T^{-1} : \mathcal{M}_n^{(0)} \to \mathcal{M}_n^{(0)}$. This concludes the proof of theorem (4.1)

We can now prove a result about inversion in an algebra of singular integral operators.

(4.7) DEFINITION. Suppose $\eta > 0$. We define $\mathcal{F}_\eta$ to be the class of all operators T with distributional kernel $K = a\delta_{x=0} + K'$ such that K' is homogeneous of degree $-n$ with mean zero on S^{n-1} and $K' \in \Lambda_{\eta^*}(S^{n-1})$ for some $\eta^* > \eta$.

Although $\mathcal{F}_\eta$ is not a Banach algebra we do have the following result.

(4.8) PROPOSITION. *Let $\eta > 0$. Then $\mathcal{F}_\eta$ is an algebra. Moreover, an element of $\mathcal{F}_\eta$ is invertible (in $\mathcal{F}_\eta$) if and only if the Fourier transform of its kernel does not vanish on the unit sphere.*
That is, an element of $\mathcal{F}_\eta$ is invertible in $\mathcal{F}_\eta$ if and only if it is invertible on $L^2(\mathbb{R}^n)$.

PROOF. Suppose S and T are elements of $\mathcal{F}_\eta$. Then, by proposition (4.6), there exists $\eta_1 > \eta$ such that $S, T \in \mathcal{H}(\frac{\eta_1}{2})$. Choose η_2 and η_3 such that $\eta < \eta_3 < \eta_2 < \eta_1$. Then S and T are each bounded from $\mathcal{M}^{(0)}_{\frac{\eta_2}{2}}$ into itself. Therefore, the same is true for their composition ST. Therefore, by theorem (4.3) of chapter 2, $ST \in \mathcal{H}(\frac{\eta_3}{2})$. Moreover, the kernel of ST is of the form $a\delta + K'$. By proposition (4.6) we have $K' \in \Lambda_{\eta_3}(S^{n-1})$ and hence $ST \in \mathcal{F}_\eta$.

The invertibility in $\mathcal{F}_\eta$ of operators with non-vanishing Fourier transform follows from the same arguments used to prove theorem (4.1). $\square$

CHAPTER 4

Maximal theorems and almost everywhere convergence of frame expansions

1. Introduction

This chapter contains three sets of theorems: equiconvergence theorems for operators of Cotlar type, almost everywhere convergence of frame expansions, and almost everywhere convergence of discrete sum approximations.

As continuous operators on spaces of molecules, Cotlar-type operators almost automatically converge in a pointwise sense on dense subspaces of classical function spaces. The classical path to almost everywhere convergence theorems is via maximal theorems, at least on those function spaces $\mathcal{B}$ where the maximal operators in question are continuous. That is, one starts with a family $\{T_\varepsilon\}_{\varepsilon>0}$ of truncated versions of the operator T, and a dense subspace $\mathcal{B}_0 \subset \mathcal{B}$ such that $T_\varepsilon g(x) \to Tg(x)$ for all x whenever $g \in \mathcal{B}_0$. Then one defines a maximal operator T_* by $T_*f(x) = \sup_\varepsilon |T_\varepsilon f(x)|$ and seeks to show that T_* is bounded on $\mathcal{B}$. If so, then a simple approximation argument shows that $T_\varepsilon f \to Tf$ converges almost everywhere for $f \in \mathcal{B}$.

Since a Cotlar-type operator T is a Calderón-Zygmund operator, one can define an associated maximal operator $\mathcal{K}_*$ by *truncation of the kernel K of T.* (see [**Ste**], p. 30). Cotlar's inequality can then be used to prove that $\mathcal{K}_*$ is weak type-(1,1) and, hence, operates continuously on L^p, $1 < p < \infty$. For operators of Cotlar type, however, there are other types of truncation that are much more natural than truncation of the kernel. For example, the frame operator

$$\mathcal{S}f(x) = \sum_{j,k} \langle f, \psi_{j,k} \rangle \psi_{j,k},$$

is a Cotlar-type operator and any *truncation via partial sums* will be uniformly bounded on molecular spaces because of uniform kernel estimates. The question of whether partial sums converge almost everywhere on L^p is typically more natural to ask than whether $\mathcal{S}_\varepsilon$ arising via truncations of the kernel

$$K(x,y) = \sum_{j,k} \psi_{j,k}(x)\overline{\psi_{j,k}(y)}$$

converge to $\mathcal{S}$. Continuity of truncation in the kernel can be used, however, as a bootstrap. In other words, suitable truncations via partial sums and truncations of the kernel are *equiconvergent* in the sense that the difference of the two associated maximal operators is bounded by the Hardy-Littlewood maximal operator. Establishing precise equiconvergence theorems for truncations of general Cotlar-type operators is the focus of the second section of this chapter. Since the Hardy-Littlewood maximal operator is continuous on L^p, $1 < p < \infty$, it will follow,

53

for example, that various types of partial sums $\mathcal{D}_N$ of Cotlar-type *discrete sum operators* $\mathcal{D}$, such as frame operators, will converge almost everywhere to $\mathcal{D}f$ for $f \in L^p$.

The third section will focus on almost everywhere convergence of frame expansions. The main observation is that the kernels associated to *truncations in scale* have all of the features of a classical approximate identity. In particular, convergence at Lebesgue points then follows from classical arguments. The ideas used here are based partially on those of Kelly, Kon and Raphael ([**KKR**]).

In the fourth section a new type of almost everywhere convergence is discussed that emphasizes continuity of the action of H on $\mathbb{R}^n$ that is so important in other features of Cotlar-type operators. It involves convergence of discrete sum operators to their continuous analogues. One consequence is the almost everywhere convergence of the normalized frame operator $\mathcal{S}_{(r,s)}$ to the identity — furnished by the Calderón reproducing formula — as the *fundamental tile* $\{(v,t) \in H : 1 \leq t \leq r,\ 0 \leq v_i \leq s,\ i = 1, \ldots, n\}$ shrinks down to the group identity $e = (0,1)$ of H, provided the shrinking is done in large enough steps. This fact gives some technical meaning to the idea that a sequence of frame operators can form an approximation to the identity.

2. Equiconvergence

To establish the almost everywhere convergence of his operator Cotlar introduced another technique that has since become basic to all maximal theorem results for singular integral operators. It will be equally basic in deriving maximal theorems for various regularizations of the $\mathcal{T}_{ab}$ operators, the most standard one arising from *truncation of the kernel*:

$$(2.1) \qquad \mathcal{K}_* : f \longrightarrow \sup_{0 < \varepsilon < \eta} \left| \mathcal{K}_{\varepsilon, \eta} f(x) \right| = \sup_{0 < \varepsilon < \eta} \left| \int_{\varepsilon < |x-y| < \eta} K(x,y) f(y)\, dy \right|.$$

Because the kernel K of $\mathcal{T}_{ab}$ satisfies the standard estimates, it follows from Cotlar's inequality that the associated maximal function $\mathcal{K}_*$ is continuous on L^p and weak-type $(1,1)$ whenever $\mathcal{T}_{ab}$ is L^2-bounded, (see, [**Ste2**], [**Mey**],vol. II); we record these facts as follows:

(2.2) COROLLARY. *For functions* $a = a(z;x), b = b(z;x)$ *in* $L^\infty(H, \mathcal{M}_\delta^{(0)}(\mathbb{R}^n))$ *let* $\mathcal{T} = \mathcal{T}_{ab}$ *and* K_t *be defined as in chapter 2. If* $\mathcal{T}_{ab}$ *is* L^2-*bounded then there is a constant* C *depending only on the dimension* n *such that for all* f:

$$\mathcal{K}_* f(x) \leq \text{const.}\, \|a\|_{\infty,\delta} \|b\|_{\infty,\delta} [M(\mathcal{T}f)(x) + Mf(x)]$$

where M *denotes the Hardy-Littlewood maximal operator. Furthermore, for all* $\alpha > 0$,

$$\left| \{ \mathcal{K}_* f(x) > \alpha \} \right| \leq \|a\|_{\infty,\delta} \|b\|_{\infty,\delta} \frac{\text{const.}}{\alpha} \int_{\mathbb{R}^n} |f(x)|\, dx.$$

For specific applications such as a.e.-convergence of discrete sum operators, different *regularizations of* $\mathcal{T}$ *over* H will be desirable. Let $\{\Omega_\varepsilon : \varepsilon > 0\}$ be a family of neighborhoods of the identity in H such that $\Omega_{\varepsilon_1} \subseteq \Omega_{\varepsilon_2}$ when $\varepsilon_1 > \varepsilon_2$

and $\bigcup_\varepsilon \Omega_\varepsilon = H$. Set

$$\mathcal{T}_{ab}^* f(x) \;=\; \sup_\varepsilon |\mathcal{T}_\varepsilon f(x)|$$

$$(2.3) \qquad = \sup_\varepsilon \left| \int_{\Omega_\varepsilon} b(z\,;x.z^{-1})\left(\frac{1}{t^n}\int_{\mathbb{R}^n} \overline{a(z\,;y.z^{-1})}f(y)\,dy\right)dz \right|$$

For suitable Ω_ε Fubini's theorem applies to the integral in (2.3) so that the *truncation in H* operator $f \to \mathcal{T}_{(\varepsilon)}f$ can be expressed as an integral kernel operator

$$(2.4) \qquad \mathcal{T}_{(\varepsilon)} : f \longrightarrow \int_{\mathbb{R}^n} \left(\int_{\Omega_\varepsilon} t^{-n}b(z\,;x.z^{-1})\overline{a(z\,;y.z^{-1})}\,dz\right)f(y)\,dy$$

solely on $\mathbb{R}^n$. Granted convergence of $\mathcal{T}_{(\varepsilon)}f(x)$ both pointwise and in norm, we can then interpret $\mathcal{T}_{ab}f$ not only as the composition $\mathcal{M}_b^*\mathcal{M}_a f$ of matrix coefficient type operators, but also as a principal value operator

$$(2.5) \qquad \mathcal{T}_{ab} : f \longrightarrow \lim_{\varepsilon \to 0} \int_{\mathbb{R}^n} \left(\int_{\Omega_\varepsilon} t^{-n}b(z\,;x.z^{-1})\overline{a(z\,;y.z^{-1})}\,dz\right)f(y)\,dy$$

as in (2.6) of chapter 2.

The simplest way of truncating in H is *truncating in dilation*,

$$(2.6) \qquad \mathcal{T}_{\varepsilon,\eta} : f \longrightarrow \int_\varepsilon^\eta \left(\int_{\mathbb{R}^n} K_t(x,y)f(y)\,dy\right)\frac{dt}{t} \;.$$

Strictly speaking, this regularization does not have the form $\mathcal{T}_{(\varepsilon)}$ of truncation in H operator, though it does when we restrict to $\eta = 1/\varepsilon$ and $\Omega_\varepsilon = \{z = (v,t) \in H : \varepsilon < t < 1/\varepsilon\}$. Other choices of Ω_ε give rise to *truncation over cones* and *truncation over hyperbolic balls* which will also be of interest.

We have seen that both L^2-boundedness as well as almost everywhere convergence of $\mathcal{T}_{ab} = \lim_{\varepsilon \to 0} \mathcal{T}_{(\varepsilon)}$ on a dense subspace of every L^p-space can be established with some conditions on a, b without any further restrictions on the Ω_ε. To establish a.e. convergence on all of L^p we adopt the usual approach of establishing the L^p-boundedness of the associated maximal operator $\mathcal{T}_{ab}^*$ by showing that different modes of truncation are *equiconvergent* in the sense that their difference is dominated by the Hardy-Littlewood maximal function. Ultimately, therefore, truncation in H is handled by comparison with truncation of the kernel and Cotlar's inequality. For some of these equiconvergence estimates we require further conditions on the decay of the functions $a = a(z\,;x), b = b(z\,;x)$, that is, on δ; nowhere, however, do we need an extra vanishing moment condition.

The first simple equiconvergence theorem compares the maximal operator associated to truncation in dilation, $\mathcal{T}_* f(x) = \sup_{0<\varepsilon<\eta} |\mathcal{T}_{\varepsilon,\eta}f(x)|$ with the truncated kernel operator, $\mathcal{K}_*$.

(2.7) **Theorem.** *For functions $a = a(z\,;x), b = b(z\,;x)$ in $L^\infty(H,\mathcal{M}_\delta^{(0)}(\mathbb{R}^n))$ the inequality*

$$\left|\{\mathcal{T}_{\varepsilon,\eta} - \mathcal{K}_{\varepsilon,\eta}\}f(x)\right| \;\leq\; \mathrm{const.}\,\|a\|_{\infty,\delta}\|b\|_{\infty,\delta}Mf(x) \qquad (\varepsilon > 0)$$

holds uniformly in $\varepsilon, \eta, a, b,$ and f.

PROOF. By definition

$$\{\mathcal{T}_{\varepsilon,\eta} - \mathcal{K}_{\varepsilon,\eta}\}f(x) = \left(\int_{|x-y|<\varepsilon} + \int_{|x-y|>\eta}\right)\int_{\varepsilon}^{\eta} K_t(x,y)\,\frac{dt}{t}f(y)\,dy$$

$$- \int_{\varepsilon<|x-y|<\eta}\left(\left(\int_0^{\varepsilon} + \int_{\eta}^{\infty}\right)K_t(x,y)\,\frac{dt}{t}\right)f(y)\,dy \ .$$

Each of these integrals has to be considered separately. First,

$$\left|\int_{|x-y|<\varepsilon}\left(\int_{\varepsilon}^{\eta} K_t(x,y)\,\frac{dt}{t}\right)f(y)\,dy\right|$$

$$\leq \ \text{const.}\,\|a\|_{\infty,\delta}\|b\|_{\infty,\delta}\int_{|x-y|<\varepsilon}\left(\int_{\varepsilon}^{\eta}\frac{t^{\delta}}{(t+|x-y|)^{n+\delta}}\,\frac{dt}{t}\right)|f(y)|\,dy$$

$$\leq \ \text{const.}\,\|a\|_{\infty,\delta}\|b\|_{\infty,\delta}\frac{1}{\varepsilon^n}\int_{|x-y|<\varepsilon}|f(y)|\,dy$$

$$\leq \ \text{const.}\,\|a\|_{\infty,\delta}\|b\|_{\infty,\delta}Mf(x)\ .$$

A similar estimate shows that

$$\left|\int_{\varepsilon<|x-y|<\eta}\left(\int_{\eta}^{\infty} K_t(x,y)\,\frac{dt}{t}\right)f(y)\,dy\right| \ \leq \ \text{const.}\,\|a\|_{\infty,\delta}\|b\|_{\infty,\delta}Mf(x)\ .$$

Estimating the remaining integrals requires a little more care. For any $\alpha, 0 < \alpha < \delta$,

$$\left|\int_{\varepsilon<|x-y|<\eta}\left(\int_0^{\varepsilon} K_t(x,y)\,\frac{dt}{t}\right)f(y)\,dy\right|$$

$$\leq \ \text{const.}\,\|a\|_{\infty,\delta}\|b\|_{\infty,\delta}\int_{\varepsilon<|x-y|<\eta}\left(\varepsilon^{\alpha}\int_0^{\varepsilon}\frac{t^{\delta-\alpha}}{(t+|x-y|)^{n+\delta}}\,\frac{dt}{t}\right)|f(y)|\,dy$$

$$\leq \ \text{const.}\,\|a\|_{\infty,\delta}\|b\|_{\infty,\delta}\int_{|y|>\varepsilon}\left(\frac{\varepsilon^{\alpha}}{|y|^{n+\alpha}}\int_0^{\varepsilon/|y|} t^{\delta-\alpha}\,\frac{dt}{t}\right)|f(x-y)|\,dy$$

$$\leq \ \text{const.}\,\|a\|_{\infty,\delta}\|b\|_{\infty,\delta}\int_{|y|>\varepsilon}\frac{\varepsilon^{\alpha}}{|y|^{n+\alpha}}|f(x-y)|\,dy$$

$$\leq \ \text{const.}\,\|a\|_{\infty,\delta}\|b\|_{\infty,\delta}Mf(x)\ ,$$

since $|y| > \varepsilon$. Similar reasoning establishes

$$\left|\int_{|x-y|>\eta}\left(\int_{\varepsilon}^{\eta} K_t(x,y)\,\frac{dt}{t}\right)f(y)\,dy\right| \ \leq \ \text{const.}\,\|a\|_{\infty,\delta}\|b\|_{\infty,\delta}Mf(x)\ .$$

Collecting together all these estimates completes the proof. $\qquad\square$

For applications to almost everywhere convergence of partial sums of the discrete sum operators the most natural maximal operator will be one in which the supremum is taken over a frustrum of a cone in H in which the aperture is allowed to increase. Correspondingly, for each $\lambda > 0$ set

$$(2.8) \qquad \Gamma_\varepsilon^\lambda = \left\{ z = (v,t) \in H : |v| \le t \ln\left(\frac{1}{\varepsilon}\right)^\lambda, \ \varepsilon < t < 1/\varepsilon \right\} .$$

The exponent λ determines the 'rate' at which the cone is opening as ε tends to zero. Now let

$$(2.9)(\text{i}) \qquad \mathcal{N}_\varepsilon^\lambda f(x) \ = \ \int_{\mathbb{R}^n} N_\varepsilon^\lambda(x,y) f(y)\, dy$$

be the integral operator whose kernel

$$(2.9)(\text{ii}) \qquad N_\varepsilon^\lambda(x,y) \ = \ \int_{\Gamma_\varepsilon^\lambda} \frac{1}{t^n} b(z\,;x.z^{-1}) \overline{a(z\,;y.z^{-1})}\, dz$$

is obtained by integration over the subset $\Gamma_\varepsilon^\lambda$ of H. As the proof of the following fundamental estimate shows, the choice of exponent λ in (2.8) is dictated by decay in the $\mathcal{M}_\delta^{(0)}$ condition.

(2.10) THEOREM. *For each $\lambda > 1 + n/\delta$ the inequality*

$$\left| \mathcal{N}_\varepsilon^\lambda f(x) \right| \ \le \ \text{const.}\, \|a\|_{\infty,\delta} \|b\|_{\infty,\delta} \left(Mf(x) + \mathcal{T}_* f(x) \right)$$

holds uniformly in x for all $a = a(z;x), b = b(z;x)$ in $L^\infty(H, \mathcal{M}_\delta^{(0)}(\mathbb{R}^n))$ and all $\varepsilon > 0$.

PROOF. The proof follows much the same pattern as that for theorem (2.7). When $|x - y|$ is very small or very large the estimate

$$(2.11)(\text{i}) \qquad \left(\int_{|x-y|<\varepsilon} + \int_{|x-y|>1/\varepsilon} \right) \left| N_\varepsilon^\lambda(x,y) f(y) \right| dy$$
$$\le \ \text{const.}\, \|a\|_{\infty,\delta} \|b\|_{\infty,\delta} Mf(x)$$

follows just as in the corresponding portion of the proof of (2.7). For the other values of $|x - y|$ however, decay becomes important. Here one compares $N_\varepsilon^\lambda(x,y)$ with the kernel

$$K^{\alpha,\beta}(x,y) \ = \ \int_\alpha^\beta \left(\int_{\mathbb{R}^n} b\left(v,t; \frac{x}{t} - v\right) \overline{a\left(v,t; \frac{y}{t} - v\right)} dv \right) \frac{dt}{t^{n+1}}$$

of the truncation in dilation operator $\mathcal{T}_{\alpha,\beta}$. The difference is estimated in terms of the radially symmetric function

$$\Phi(x) \ = \ \frac{1}{1 + |x|^n} \left(\frac{1}{1 + \ln(1 + |x|)} \right)^{\frac{\lambda\delta}{n+\delta}} .$$

The choice of λ ensures that Φ is also integrable, so that

$$(2.11)(\text{ii}) \qquad \sup_{\varepsilon>0} \left| \Phi_\varepsilon * f(x) \right| \ \le \ \text{const.}\, Mf(x) .$$

The specific scales α and β depend on ε and on the relative positions of x and y as dictated by the following lemma. For convenience of notation set $h = 1 + \ln\left(\frac{1}{\varepsilon}\right)^\lambda$.

(2.12) LEMMA. *When $\varepsilon < |x - y| < 1/\varepsilon$:*
(i) *the inequality*

$$\left| N_\varepsilon^\lambda(x, y) - K_{|x|/h, 1/\varepsilon}(x, y) \right| \leq \text{const.} \, \|a\|_{\infty, \delta} \|b\|_{\infty, \delta} \Phi_\varepsilon(x - y)$$

holds uniformly in x and y provided $\varepsilon < |x|/h < 1/\varepsilon$;
(ii) *the inequality*

$$\left| N_\varepsilon^\lambda(x, y) - K_{\varepsilon, 1/\varepsilon}(x, y) \right| \leq \text{const.} \, \|a\|_{\infty, \delta} \|b\|_{\infty, \delta} \Phi_\varepsilon(x - y)$$

holds uniformly in x and y provided $|x|/h < \varepsilon$; and
(iii) *the inequality*

$$\left| N_\varepsilon^\lambda(x, y) \right| \leq \text{const.} \, \|a\|_{\infty, \delta} \|b\|_{\infty, \delta} \Phi_\varepsilon(x - y)$$

holds uniformly in x and y provided $1/\varepsilon < |x|/h$.

Once these estimates have been established (2.10) follows immediately using both parts of (2.11). $\qquad\square$

PROOF. We shall only prove (i) as (ii) and (iii) follows by simpler variants of the same argument. To prove (i) suppose first that

$$(2.13) \qquad\qquad |x - y| \leq h^{r-1}|x| \qquad (r = 1/(\delta + n)) .$$

Therefore,

$$N_\varepsilon^\lambda(x, y) - K_{|x|/h, 1/\varepsilon}(x, y)$$

$$= \left(\int_\varepsilon^{|x|/h} \int_{|v| < t \ln(1/\varepsilon)^\lambda} - \int_{|x|/h}^{1/\varepsilon} \frac{1}{t^n} \int_{|v| > t \ln(1/\varepsilon)^\lambda} \right) b(z \, ; x.z^{-1}) \overline{a(z \, ; y.z^{-1})} \, dz .$$

Consequently

$$\left| N_\varepsilon^\lambda(x, y) - K_{|x|/h, 1/\varepsilon}(x, y) \right| \leq \text{const.} \, \|a\|_{\infty, \delta} \|b\|_{\infty, \delta} \left(I_1 + I_2 \right)$$

where

$$I_1 = \int_\varepsilon^{|x|/h} \frac{1}{t^n} \int_{|v| < t \ln(1/\varepsilon)^\lambda} \frac{1}{\left(1 + |x.z^{-1}|\right)^{n+\delta}} \frac{1}{\left(1 + |y.z^{-1}|\right)^{n+\delta}} \, dz$$

and

$$I_2 = \int_{|x|/h}^{1/\varepsilon} \frac{1}{t^n} \int_{|v| > t \ln(1/\varepsilon)^\lambda} \frac{1}{\left(2 + |x.z^{-1}|\right)^{n+\delta}} \frac{1}{\left(1 + |y.z^{-1}|\right)^{n+\delta}} \, dz .$$

The restrictions on v and t ensure that in I_1 the inequality

$$1 + |x.z^{-1}| \geq 1 + \frac{|x|}{t} - \ln(1/\varepsilon)^\lambda \geq 2$$

holds uniformly, while in I_2 the corresponding inequality is

$$2 + |x.z^{-1}| \geq 2 + \ln(1/\varepsilon)^\lambda - \frac{|x|}{t} \geq 1 .$$

Thus

$$I_1 \leq \int_\varepsilon^{|x|/h} \frac{1}{\left(1 + \frac{|x|}{t} - h\right)^{n+\delta}} \frac{dt}{t^{n+1}} \leq \frac{\text{const.}}{|x|^n} \int_h^\infty \frac{s^{n-1}}{\left(1 + s - h\right)^{n+\delta}} \, ds .$$

Similarly,

$$I_2 \;\leq\; \int_{|x|/h}^{1/\varepsilon} \frac{1}{(1 + h - \frac{|x|}{t})^{n+\delta}} \, \frac{dt}{t^{n+1}} \;\leq\; \frac{\text{const.}}{|x|^n} \int_0^h \frac{s^{n-1}}{(1 + h - s)^{n+\delta}} \, ds \; .$$

In view of (6.14) and the restriction $\varepsilon < |x - y| < 1/\varepsilon$,

$$I_1 + I_2 \;\leq\; \text{const.}\, \frac{h^{n-1}}{|x|^n} \;\leq\; \text{const.}\, \frac{1}{|x - y|^n} \left(\frac{1}{h}\right)^{\frac{\delta}{n+\delta}} \;\leq\; \text{const.}\, \Phi_\varepsilon(x - y) \; .$$

Consequently, the inequality in (i) holds for the range of x, y specified by (2.13).
Next suppose that

$$(2.14) \qquad\qquad |x - y| \;\geq\; h^{r-1}|x| \qquad (r = 1/(\delta + n)) \; ,$$

i.e., $\varepsilon < |x|/h \leq |x - y|/h^r$. Then

$$N_\varepsilon^\lambda(x, y) - K_{|x|/h,1/\varepsilon}(x, y) = \left(\int_\varepsilon^{|x|/h} \int_{|v| < t \ln(1/\varepsilon)^\lambda} - \int_{|x|/h}^{|x-y|/h^r} \int_{|v| > t \ln(1/\varepsilon)^\lambda} \right.$$

$$\left. - \int_{|x-y|/h^r}^{1/\varepsilon} \int_{|v| > t \ln(1/\varepsilon)^\lambda} \right) t^{-2n} b(z\,;\, x.z^{-1}) \overline{a(z\,;\, y.z^{-1})} \, dz \; .$$

Consequently

$$\left| N_\varepsilon^\lambda(x, y) - K_{|x|/h,1/\varepsilon}(x, y) \right| \;\leq\; \text{const.}\, \|a\|_{\infty,\delta} \|b\|_{\infty,\delta} \left(I_3 + I_4 \right)$$

where

$$I_3 \;=\; \int_\varepsilon^{|x-y|/h^r} \frac{1}{t^n} \int_{\mathbb{R}^n} \frac{1}{\left(1 + |x.z^{-1}|\right)^{n+\delta}} \frac{1}{\left(1 + |y.z^{-1}|\right)^{n+\delta}} \, dz$$

and

$$I_4 \;=\; \int_{|x-y|/h^r}^{1/\varepsilon} \frac{1}{t^n} \int_{|v| > t \ln(1/\varepsilon)^\lambda} \frac{1}{\left(2 + |x.z^{-1}|\right)^{n+\delta}} \frac{1}{\left(1 + |y.z^{-1}|\right)^{n+\delta}} \, dz \; .$$

By Basic Estimate A(i)

$$I_3 \;\leq\; \text{const.} \int_\varepsilon^{|x-y|/h^r} \frac{t^\delta}{(t + |x - y|)^{n+\delta}} \frac{dt}{t}$$

$$\leq\; \frac{\text{const.}}{(\varepsilon + |x - y|)^{n+\delta}} \int_\varepsilon^{|x-y|/h^r} t^\delta \frac{dt}{t} \;\leq\; \text{const.}\, \frac{h^{-r\delta}}{(\varepsilon + |x - y|)^n} \; .$$

$$\leq\; \frac{\text{const.}}{(\varepsilon + |x - y|)^{n+\delta}} \left(\frac{1}{1 + \ln(1/\varepsilon)}\right)^{\frac{\lambda\delta}{n+\delta}} \;\leq\; \text{const.}\, \Phi_\varepsilon(x - y)$$

On the other hand, since $t \geq |x|/h$ in I_4, the restrictions on v and t ensure that

$$2 + |x.z^{-1}| \;\geq\; 2 + \ln(1/\varepsilon)^\lambda - h = 1$$

holds uniformly in I_4. Thus

$$I_4 \;\leq\; \frac{\text{const.}}{(1+h)^{n+\delta}} \int_{|x-y|/h^r}^{1/\varepsilon} \frac{dt}{t^{n+1}}$$

$$\leq\; \text{const.} \frac{h^{n(r-1)-\delta}}{|x-y|^n} \;\leq\; \frac{\text{const.}\, h^{nr-1}}{(\varepsilon+|x-y|)^{n+\delta}} \;\leq\; \text{const.}\,\Phi_\varepsilon(x-y)\;,$$

since $\varepsilon < |x-y| < 1/\varepsilon$ on $|x-y|$. Consequently, the inequality in (i) also holds for
the range of x, y specified by (2.14), and therefore for all $\varepsilon < |x-y|/h < 1/\varepsilon$. $\square$

As a final maximal theorem for operators of Cotlar type we consider truncation
in H by hyperbolic balls *centered at* x. A hyperbolic ball $B_h(r)$ centered at the
identity in H is a Euclidean ball in $H = \mathbb{R}_+^{n+1}$ centered at the point $(0, \cosh r)$
having radius $\sinh r$. We also denote by $B_h(x; r)$ the right translate $B_h(r) \cdot (x, 0)$.
Let $\mathcal{B}_\varepsilon$ denote the hyperbolic centered truncation operator,

$$(2.15)(\mathrm{i}) \qquad \mathcal{B}_\varepsilon : f \mapsto \int_{\mathbb{R}^n} \left(\int_{B_h(x;\log(\varepsilon))} t^{-n} b(z; x \cdot z^{-1}) \overline{a(z; y \cdot z^{-1})} dz \right) f(y)\, dy$$

Then we have also the equiconvergence result:

(2.16) THEOREM. *For functions* $a = a(z; x), b = b(z; x)$ *in* $L^\infty(H, \mathcal{M}_\delta^{(0)}(\mathbb{R}^n))$
the inequality

$$|\{\mathcal{B}_\varepsilon - \mathcal{K}_\varepsilon\} f(x)| \;\leq\; \text{const.}\, \|a\|_{\infty,\delta} \|b\|_{\infty,\delta} M f(x) \qquad (\varepsilon > 0)$$

holds uniformly in x, ε, a, b.

The proof involves introducing in addition a *rectangular* truncation operator
where one truncates over the set $R_\varepsilon(x) = \{(v, t) \in H : |x-v| < 1/\varepsilon, \quad \varepsilon < t < 1/\varepsilon\}$.
A simple computation shows that the corresponding *inscribed* and *circumscribed*
rectangles have bounded hyperbolic volume ratios. Loosely speaking, this means
that a maximal operator arising from truncation over hyperbolic balls is equivalent
to the one arising via truncation over *centered rectangles*. The latter is easily
comparable to the truncation in dilation operator by the techniques used above.
The $\mathcal{B}_\varepsilon$ operators are natural truncations in situations that emphasize hyperbolic
geometry.

These equiconvergence results can be simplified in the homogeneous case. Given
functions $\phi = \phi(x)$, $\psi = \psi(x)$, $\Psi_t = \psi_t * \tilde{\phi}_t$, and $\Psi = \int_0^\infty \Psi_t \frac{dt}{t}$, the truncation in
dilation and truncation in space operators have the forms

$$(2.15)(\mathrm{ii})\; \mathcal{E}_\varepsilon : f \longrightarrow \int_\varepsilon^{1/\varepsilon} \Psi_t * f \frac{dt}{t}; \qquad \mathcal{F}_\varepsilon : f \longrightarrow \int_{\varepsilon<|x-y|<1/\varepsilon} \Psi(x-y) f(y)\, dy\;.$$

Notice that when $\phi(x) = \psi(x)$ is the function

$$\phi(x) \;=\; \left. \frac{\partial P_t(x)}{\partial t} \right|_{t=1}$$

then the operator $\mathcal{E}_*$ will essentially be the usual nontangential maximal function.

(2.17) THEOREM. *When $\phi = \phi(x)$ and $\psi = \psi(x)$ are functions in $\mathcal{M}_\delta^{(0)}(\mathbb{R}^n))$ and the associated truncation operators $\mathcal{E}_\varepsilon, \mathcal{F}_\varepsilon$ are defined by (2.16), the limit*

$$\lim_{\varepsilon \to 0} \{\mathcal{E}_\varepsilon - \mathcal{F}_\varepsilon\} f(x) \;=\; -\left(\int_{\mathbb{R}^n} \psi * \phi^*(y) \log(|y|)\, dy \right) f(x)$$

exists both pointwise for any continuous compactly supported f and in L^1-norm for any f in $L^1(\mathbb{R}^n)$.

The proof follows from a change to polar coordinates and Fubini's theorem. We omit the details.

Next we show how theorem (2.16) can be used to prove a.e. convergence results for certain partial sums of discrete sum operators.

Recall that we set, for each fixed choice of $r > 1$ and $s > 0$,

$$\lambda_{j,k} \;=\; (sk/r^j, 1/r^j) \qquad \left(k \in \mathbb{Z}^n, j \in \mathbb{Z} \right)$$

$$\Delta \;=\; \Delta^{(r,s)} \;=\; \{ (y,t) \in H : 0 \le y_\ell < s\,, 1 \le t < r, \} \qquad (y = (y_1, \dots, y_n))\,.$$

We also noted that the right translates $\Delta_{j,k} = \Delta.\lambda_{j,k}$ of Δ provide an affine tiling of H, with $|\Delta_{j,k}| = \frac{s^n}{n}\left(1 - \frac{1}{r^n}\right) = |\Delta|$ independently of j and k. Now to each pair of functions ϕ, ψ in $\mathcal{M}_\delta^{(0)}(\mathbb{R}^n)$ we associate the functions

(2.18)(i) $$\qquad a(z\,;y) \;=\; \frac{1}{|\Delta|^{1/2}} \sum_{j,k} \chi_\Delta(z\lambda_{j,k}^{-1}) U(z\lambda_{j,k}^{-1})\phi(y)$$

and

(2.18)(ii) $$\qquad b(z\,;x) \;=\; \frac{1}{|\Delta|^{1/2}} \sum_{j,k} \chi_\Delta(z\lambda_{j,k}^{-1}) U(z\lambda_{j,k}^{-1})\psi(x)\,.$$

where χ_Δ is the characteristic function of Δ and $U(w)$ is defined as in chapter 1. Since the $\Delta_{j,k}$ tile H,

$$\|a(z\,;.)\|_{\mathcal{M}_\delta} \;\le\; \frac{1}{|\Delta|^{1/2}} \sup_{w \in \Delta} \|U(w)\phi\|_{\mathcal{M}_\delta}\,.$$

On the other hand, since U is bounded on compact sets of H,

$$\sup_{w \in \Delta} \|U(w)\phi\|_{\mathcal{M}_\delta} \;\le\; \text{const.}\,\|\phi\|_{\mathcal{M}_\delta}$$

for some constant depending on Δ. Thus

(2.19) $$\qquad \|a\|_{\infty,\delta} \;\le\; C_\Delta \|\phi\|_{\mathcal{M}_\delta} < \infty$$

and there is a similar estimate for $b(z\,;x)$. Thus a,b are L^∞-bounded on H and

(2.20) $$\quad \frac{1}{t^n} b(z\,;x.z^{-1}) \overline{a(z\,;y.z^{-1})} \;=\; \frac{1}{|\Delta|} \sum_{j,k} \chi(z\lambda_{j,k}^{-1}) r^{jn} \psi(r^j x - sk) \overline{\phi(r^j y - sk)}\,.$$

The corresponding $\mathcal{T}_{ab}$-operator is just the frame operator

$$\mathcal{D}_{\phi\psi}(f)(x) \;=\; \sum_{j,k} r^{jn}\Big(\int_{\mathbb{R}^n} f(y)\overline{\phi(r^j y - sk)}\,dy \Big)\psi(r^j x - sk)$$

$$(2.21) \qquad\qquad = \sum_{j,k} \langle f, \phi_{j,k}\rangle \psi_{j,k}(x)$$

associated with ϕ, ψ. It is natural to call this a *Discrete Sum Operator* or, more precisely, a *Discrete Sum Operator having mesh size* (r, s) to emphasize its dependence on the affine lattice $\{\lambda_{j,k}\}$.

(2.22) COROLLARY. *The Discrete sum operator $\mathcal{D}_{\phi\psi}$ is bounded on $L^2(\mathbb{R}^n)$ for each choice of r, s whenever both ϕ and ψ are functions in $\mathcal{M}_\delta^{(0)}(\mathbb{R}^n)$. If one but not both of the generating functions ϕ and ψ fails to have vanishing moment, the corresponding $\mathcal{D}_{\phi\psi}$ is still Restrictedly Bounded.*

Next we closely investigate the implications of the various equiconvergence results above to almost everywhere convergence of partial sums of discrete sum operators. In particular, we are concerned with a.e. convergence of the partial sums of $\mathcal{D}_{\phi\psi}$. Of course convergence will be convergence to the frame operator and not to the identity, cf., section 3 below.

(2.23) THEOREM (CONVERGENCE OF DISCRETE PARTIAL SUMS). *Let $\mathcal{S} = \mathcal{D}_{\phi\psi}$ be as in (2.21) where both $\phi, \psi \in \mathcal{M}_\delta^{(0)}$ for some $\delta > 0$. For $N > 0$ we define the N-th partial sum operator of $\mathcal{S}$ of opening λ to be:*

$$\mathcal{S}_N^\lambda : f \longrightarrow \frac{1}{|\Delta|} \sum_{|j| \leq N, |k_1|,\dots|k_n| \leq N^\lambda} r^{jn}\psi(r^j x - sk)\Big(\int_{\mathbb{R}^n} \overline{\phi(r^j y - sk)}f(y)\,dy \Big)$$

where $\lambda > 1 + \frac{n}{\delta}$. Then $\mathcal{S}_N^\lambda$ converges to $\mathcal{S}$ as $N \to \infty$, both in $\mathcal{L}(L^p)$ and almost everywhere.

PROOF. To prove the result we must show that $\mathcal{S}_N^\lambda$ is *essentially* an $\mathcal{N}_\varepsilon^\lambda$-type operator where ε corresponds to N. This done, we can then invoke the equiconvergence result (2.10) for truncated cones together with the convergence of $\mathcal{S}$ to obtain the boundedness of

$$\mathcal{S}_*^\lambda f(x) \;=\; \sup_{N>0} |\mathcal{S}_N^\lambda f(x)|$$

and consequently the proof of the theorem. Thus we need to show that the kernel of $\mathcal{S}_N^\lambda$ approximately has the form

$$(2.24) \qquad\qquad \int_{\Gamma_\varepsilon^\lambda} \frac{1}{t^n} b(z\,;x.z^{-1})\overline{a(z\,;y.z^{-1})}\,dz$$

where, as before,

$$\Gamma_\varepsilon^\lambda \;=\; \Big\{(v,t) \in H : |v| \leq t \ln(\frac{1}{\varepsilon})^\lambda,\ \varepsilon < t < 1/\varepsilon \Big\}\,.$$

The first step is to notice that, by a routine calculation

$$(2.25) \qquad \mathcal{S}_N : f \longrightarrow \sum_{|j|<N, k\in\mathbb{Z}^n} r^{jn}\Big(\int_{\mathbb{R}^n} f(y)\overline{\phi(r^j y - sk)}\,dy \Big)\psi(r^j x - sk)\,,$$

has the form

$$f \longrightarrow \int_{r^{-N}}^{r^N} \left(\int_{\mathbb{R}^n} \left(\int_{\mathbb{R}^n} b(v,t; \tfrac{x-v}{t}) \overline{a(v,t; \tfrac{x-v}{t})} dv \right) f(y)\, dy \right) \frac{dt}{t^{2n+1}}$$

where a, b are as in (2.18). So $\mathcal{S}_N$ is just a truncation in scale where $\varepsilon = r^{-N}$.

The next step is to note that $v \in \Gamma_\varepsilon^\lambda \equiv \Gamma_N^\lambda$, that is $|v| < t \ln(\tfrac{1}{\varepsilon})^\lambda$, means that $|v| < t \ln(r^N)^\lambda = tr'N^\lambda$ where $r' = \ln(r)^\lambda$. In view of (2.20) the question is, when $z = (v,t)$ with $t \in (r^{-N}, r^N)$ and $v \in \Gamma_N^\lambda$, which k force $\chi_\Delta(z\lambda_{j,k}^{-1}) = 0$? The sum in $\mathcal{S}_N^\lambda$ will be restricted to those remaining k.

Now fix $t \in [2^{-N}, 2^N)$ If $(r^j v - sk, r^j t) \in \Delta$ then

$$\frac{k_\ell s}{r^j} \leq v_\ell < \frac{(k_\ell + 1)s}{r^j}, \qquad (1 \leq \ell \leq n) \ .$$

Choosing j so that $1 \leq r^j t \leq r$ then gives

$$\frac{k_\ell s t}{r} \leq v_\ell < \frac{(k_\ell + 1)s t}{r^j} \ .$$

Therefore, in order that $z = (v,t)$ satisfy $\chi_\Delta(z\lambda_{j,k}^{-1}) \neq 0$ we need

$$|v_\ell| \leq r' t N^\lambda \quad \text{and} \quad \frac{|k_\ell| s t}{r} \leq |v_\ell| < (|k_\ell| + 1)st \ .$$

This forces $|k_\ell| \leq \frac{rr'}{s} N^\lambda$, that is, $|k_\ell| \leq cN^\lambda$. This verifies that, up to insignificant geometrical factors, $\mathcal{S}_N^\lambda$ is an $\mathcal{N}_\varepsilon^\lambda$-type operator. This proves the theorem. $\square$

3. Almost everywhere convergence of wavelet frame expansions

Suppose ψ is an orthonormal wavelet and $f \in L^2(\mathbb{R})$. Then the wavelet expansion

$$f = \sum_{j \in \mathbb{Z}} \sum_{k \in \mathbb{Z}} \langle f, \psi_{j,k} \rangle \psi_{j,k}$$

converges unconditionally in L^2-norm. For well-behaved ψ one can establish pointwise convergence of this sum ([**KKR**]). We will be concerned with the pointwise convergence of wavelet frames. The partial sums taken in this section will be less restrictive than in previous sections. However, the results apply only when the underlying operator is the identity as expressed by the frame reproducing formula. The regularity conditions we will need are stated in terms of molecules. In this section we shall assume for notational convenience that $\psi_{j,k}(x) = 2^{nj/2}\psi(2^j x - k)$. Frames arising from different scaling factors and translation steps can be dealt with in the same way.

Let $\psi \in \mathcal{M}_\delta^{(0)}$ and suppose that $\{\psi_{j,k}\}$ forms a dyadic frame for $L^2(\mathbb{R}^n)$. Let $\{\rho_{j,k}\}$ denote the dual frame. We have seen that the $\rho_{j,k}$'s typically are not translations and dilations of a single function ρ. However, for fixed k, the functions $\rho_{j,k}$ are dilations of the single function $\rho_{0,k}$. That is, $\rho_{j,k}(x) = 2^{\frac{nj}{2}} \rho_{0,k}(2^j x)$. This follows from the fact that the frame operator $\mathcal{D}$, and hence $\mathcal{D}^{-1}$, commutes with dyadic dilations (see [**HeW**]). This fact will be fundamental in our convergence proofs.

We need to assume some regularity on the dual frame. Suppose that $\rho_{0,k} \in \mathcal{M}_\delta^{(0)}(k,1)$ with norm bounded independent of $k \in \mathbb{Z}^n$. Since $\rho_{j,k}$ is just a dilation of $\rho_{0,k}$ we can obtain similar molecular estimates for all of the $\rho_{j,k}$'s. For the frames constructed in this paper, the frame operator was inverted on molecular spaces. Therefore the dual frames satisfy the necessary estimates.

(3.1) THEOREM. *Suppose $\{\psi_{j,k}\}$ is a frame for $L^2(\mathbb{R}^n)$ with dual frame $\{\rho_{j,k}\}$ as above. Then for $f \in L^p(\mathbb{R}^n)$,*

$$f = \sum_{j\in\mathbb{Z}} \sum_{k\in\mathbb{Z}^n} \langle f, \psi_{j,k}\rangle \rho_{j,k}$$

$$= \sum_{j\in\mathbb{Z}} \sum_{k\in\mathbb{Z}^n} \langle f, \rho_{j,k}\rangle \psi_{j,k}$$

where the series converges pointwise at every Lebesgue point of f.

We can express the identity operator as

$$If = \sum_{j\in\mathbb{Z}} \sum_{k\in\mathbb{Z}^n} \langle f, \psi_{j,k}\rangle \rho_{j,k}$$

with convergence in $L^2(\mathbb{R}^n)$, say. For sets $\mathcal{J} \subset \mathbb{Z}$ and $\mathcal{K} \subset \mathbb{Z}^n$ (not necessarily finite) define

$$I_{\mathcal{J},\mathcal{K}} f = \sum_{j\in\mathcal{J}} \sum_{k\in\mathcal{K}} \langle f, \psi_{j,k}\rangle \rho_{j,k}$$

and let

$$K_{\mathcal{J},\mathcal{K}}(x,y) = \sum_{j\in\mathcal{J}} \sum_{k\in\mathcal{K}} \rho_{j,k}(x)\,\overline{\psi_{j,k}(y)}$$

denote the kernel of $I_{\mathcal{J},\mathcal{K}}$. We know from theorem (3.1) of chapter 2 that the kernels $K_{\mathcal{J},\mathcal{K}}(x,y)$ satisfy standard estimates independent of the sets $\mathcal{J}$ and $\mathcal{K}$. Therefore the operators $I_{\mathcal{J},\mathcal{K}}$ are uniformly bounded on molecule spaces and other function spaces independent of $\mathcal{J}$ and $\mathcal{K}$.

We now study the kernel K,

$$K(x,y) = \sum_{j\in\mathbb{Z}} \sum_{k\in\mathbb{Z}^n} \rho_{j,k}(x)\overline{\psi_{j,k}(y)} \ ,$$

of I more closely. To do this we look at the inner sum first. Define

$$\Psi(x,y) = \sum_{k\in\mathbb{Z}^n} \rho_{0,k}(x)\overline{\psi_{0,k}(y)}$$

$$= \sum_{k\in\mathbb{Z}^n} \rho_{0,k}(x)\overline{\psi(y-k)} \ .$$

Since $\mathcal{D}^{-1}$ commutes with dyadic dilations we can write

$$2^{nj}\Psi(2^j x, 2^j y) = 2^{nj} \sum_{k\in\mathbb{Z}^n} \rho_{0,k}(2^j x)\overline{\psi_{0,k}(2^j y)}$$

$$= \sum_{k\in\mathbb{Z}^n} \rho_{j,k}(x)\overline{\psi_{j,k}(y)} \ .$$

Therefore we can write $K(x,y) = \sum_{j\in\mathbb{Z}} 2^{nj}\Psi(2^j x, 2^j y)$. Using the size estimates of the molecules $\psi_{0,k}$ and $\rho_{0,k}$ and a discrete form of Basic Estimate A(i) we obtain

$$(3.2) \qquad |\Psi(x,y)| \le \frac{\text{const.}}{(1+|x-y|)^{n+\delta}} \ .$$

whenever $\delta < \gamma$. Define the operators

$$Pf(x) \;=\; \sum_{j<0}\sum_{k\in\mathbb{Z}^n} \langle f, \psi_{j,k}\rangle \rho_{j,k} \;=\; \int \left\{ \sum_{j<0}\sum_{k\in\mathbb{Z}^n} \rho_{j,k}(x)\overline{\psi_{j,k}(y)} \right\} f(y)\,dy$$

$$Qf(x) \;=\; \sum_{j\geq 0}\sum_{k\in\mathbb{Z}^n} \langle f, \psi_{j,k}\rangle \rho_{j,k} \;=\; \int \left\{ \sum_{j\geq 0}\sum_{k\in\mathbb{Z}^n} \rho_{j,k}(x)\overline{\psi_{j,k}(y)} \right\} f(y)\,dy.$$

Denote the kernels of these operators by $P(x,y)$ and $Q(x,y)$, respectively. We can then write

$$P(x,y) \;=\; \sum_{j<0} 2^{nj}\Psi(2^j x, 2^j y)$$

$$Q(x,y) \;=\; \sum_{j\geq 0} 2^{nj}\Psi(2^j x, 2^j y).$$

Since $f = \sum_{j,k}\langle f, \psi_{j,k}\rangle \rho_{j,k}$ we know that $K(x,y) = P(x,y) + Q(x,y) = \delta(x - y)$, where δ is the Dirac mass. Hence, at least formally, $P + Q = 0$ away from the diagonal. The following theorem is proved in [**KKR**] for orthonormal wavelets. The proofs carry over without incidence to the present case.

(3.3) THEOREM [**KKR**]. *The sum for $P(x,y)$ converges uniformly to a continuous function. The sum for $Q(x,y)$ converges uniformly on sets with positive distance from the diagonal. Moreover,*

$$P(x,y) \;+\; Q(x,y) \;=\; 0, \qquad (x \neq y)\,.$$

The size estimate (3.2) on Ψ gives the following two estimates

$$|Q(x,y)| \;\leq\; \text{const.}\sum_{j\geq 0} \frac{2^{-j\delta}}{(2^{-j} + |x - y|)^{n+\delta}} \;\leq\; \frac{\text{const.}}{|x - y|^{n+\delta}}\sum_{j\geq 0} 2^{-j\delta} \;\leq\; \frac{\text{const.}}{|x - y|^{n+\delta}}$$

$$|P(x,y)| \;\leq\; \text{const.}\sum_{j<0} \frac{2^{-j\delta}}{(2^{-j} + |x - y|)^{n+\delta}} \;\leq\; \text{const.}\sum_{j<0} 2^{nj} \;\leq\; \text{const.}$$

Combining these two estimates along with theorem (3.3) we obtain the estimate

$$(3.4) \qquad\qquad |P(x,y)| \;\leq\; \frac{\text{const.}}{(1 + |x - y|)^{n+\delta}}\,.$$

Thus we have a radially decreasing L^1-majorant for the kernel P.

Define an operator (analogous to the projection onto V_N in wavelet theory) $P_N f = \sum_{j<N}\sum_{k\in\mathbb{Z}^n}\langle f, \psi_{j,k}\rangle \rho_{j,k}$. Then the kernel of P_N is

$$(3.5) \qquad P_N(x,y) \;=\; \sum_{j<N} 2^{nj}\Psi(2^j x, 2^j y) \;=\; 2^{Nn}P(2^N x, 2^N y)\,.$$

This equation (3.5) and the majorization (3.4) suggest a resemblance to a classical convolution approximation to the identity. To complete the analogy we need to show that the function P has integral one.

(3.6) LEMMA. $\int P(x,y)\,dy = 1$ *for every* $x \in \mathbb{R}^n$ *and* $\int P(x,y)\,dx = 1$ *for every* $y \in \mathbb{R}^n$.

PROOF. Let $\alpha(x) = \int P(x, y)\, dy$. We want to show that $\alpha(x) \equiv 1$. It follows from the inequality in (3.4) that α is a continuous function.

We first claim that $\alpha(2x) = \alpha(x)$. It is not hard to see that $P(x, y) + \Psi(x, y) = 2^n P(2x, 2y)$ since each side of this equation is equal to $\sum_{j \leq 0} 2^{nj} \Psi(2^j x, 2^j y)$. Integrating this identity with respect to y, along with a change of variables, establishes the claim since $\int \Psi(x, y)\, dy = 0$. Thus $\alpha(2^N x) = \alpha(x)$ for every $N \in \mathbb{Z}$.

It now follows that $\alpha(x)$ is a constant function. Indeed, for fixed x, we can let $N \to -\infty$ in the identity $\alpha(2^N x) = \alpha(x)$ to see that $\alpha(x) = \alpha(0)$ (recall that α is a continuous function). From classical approximation to the identity arguments it follows that $P_N f \to \alpha f$ for every $f \in \mathcal{S}$. We know, however, that $P_N f \to f$ in $L^2(\mathbb{R}^n)$. Therefore, α must be identically 1.

To show that $\int P(x, y)\, dx = 1$ we repeat the previous arguments but use the identity $f = \sum_{j,k} \langle f, \rho_{j,k} \rangle \psi_{j,k}$ instead of $f = \sum_{j,k} \langle f, \psi_{j,k} \rangle \rho_{j,k}$. $\square$

The following result now follows from classical approximation to the identity arguments.

(3.7) THEOREM. *$P_N f$ converges to f in $L^p(\mathbb{R}^n)$, for $f \in L^p(\mathbb{R}^n)$. $P_N f(x)$ converges to $f(x)$ for every Lebesgue point x of f.*

4. Affine approximations to the identity

Our goal here is to prove a maximal theorem for frame operators where the supremum is taken pointwise over a set of operators with increasingly fine lattice density determined by a sequence of fundamental tiles $\Delta^{(\ell)}$ that shrinks to $(0, 1) \in H$. The mesh size should decrease in 'jumps' at a certain rate in order to prove a desired weak-type $(1,1)$ estimate for the maximal operator. Such a 'lacunarity' condition is imposed, for example, when the fundamental tiles are nested, meaning that $\Delta^{(\ell)}$ is a union of translates of $\Delta^{(\ell+1)}$. It is unknown whether such a maximal theorem is possible when Δ shrinks continuously to $(0, 1)$. In view of the nature of these questions we shall assume *throughout the section* that $0 < \delta < 1$.

In what follows, fix $\phi, \psi \in \mathcal{M}_\delta^{(0)}$. As before, let $\mathcal{T}_{\phi, \psi}$ denote the operator

$$f \mapsto \int_H \langle f, U^*(z)\phi \rangle U^*(z)\psi\, dz$$

and for a fixed mesh of size (r, s) let $\mathcal{S}_{\phi\psi}$ be the corresponding discrete sum operator. Actually, we shall be concerned with a sequence $\mathcal{S}_{\phi\psi}^{(\ell)}$ of such operators where the fundamental tiles $\Delta^{(\ell)}$ have decreasing mesh size (r_ℓ, s_ℓ). Then we can define a *maximal operator in mesh size*

$$\mathcal{S}_* f(x) = \sup_\ell |\mathcal{S}_{\phi\psi}^{(\ell)} f(x)|.$$

Boundedness of such an operator will imply almost everywhere convergence of the sequence $\mathcal{S}_{\phi\psi}^{(\ell)} f(x)$ to $\mathcal{T}_{\phi, \psi} f(x)$ as the mesh size tends to zero. In order to state a precise condition on mesh shrinkage to guarantee such boundedness, we first state a precise corollary of the proof of theorem (1.1) of chapter 3:

(4.1) COROLLARY. *Let $\phi \in \mathcal{M}_\delta(\mathbb{R}^n)$ and let $w \in \Delta(r, s)$. Then for $0 < \gamma < \delta$,*

$$\|(I - U(w))\phi\|_{\mathcal{M}_\gamma} \leq \mathrm{const.}[(r - 1)^\delta + s^\delta]^{1 - \gamma/\delta}\|\phi\|_{\mathcal{M}_\delta}$$

where the constant depends on n, γ, δ but not r, s or ϕ.

It is the term $[(r-1)^\delta + s^\delta]^{1-\gamma/\delta}$ that will be crucial here:

(4.2) THEOREM. *Let $\Delta^{(\ell)}$ be a decreasing family of fundamental tiles with mesh sizes (r_ℓ, s_ℓ), and suppose that for some $0 < \gamma < \delta$, $\sum_{\ell=1}^{\infty} [(r_\ell - 1)^\delta + s_\ell^\delta]^{1-\gamma/\delta}$ converges. Then the maximal operator S_* is bounded on L^p, $1 < p < \infty$. In particular, for each $f \in L^p$, $S_{\phi\psi}^{(\ell)} f$ converges to $T_{\phi\psi} f$ both in L^p and almost everywhere.*

The almost everywhere convergence is an immediate corollary of L^p boundedness of S_* together with pointwise convergence on $\mathcal{M}_\gamma$.

One simple example corresponds to the case where one builds a sequence of nested lattices by halving the mesh size of successive lattices both in space and logarithmically in scale. In other words, let $\Delta^{(\ell)}$ be the fundamental domain having mesh size $(r_\ell, s_\ell) = (2^{1/2^\ell}, 1/2^\ell)$. It is easy to see that the corresponding lattices $\Lambda^{(\ell)} = \{\lambda_{j,k}^{(\ell)}\}$ are nested. Furthermore, since $2^{1/2^\ell} - 1 \le C2^{-\ell}$ it follows that $[(r_\ell - 1)^\delta + s_\ell^\delta]^{1-\gamma/\delta} \le C2^\ell(\gamma - \delta)$ so that the condition of Theorem (4.2) will be satisfied for any $\gamma < \delta$. Therefore the criterion for convergence in mesh size holds for this sequence of lattices.

It is time to say something about the proof. The main idea is to get control on the family of *error operators* $\mathcal{E}_{\phi\psi}^{(\ell)} = T_{\phi\psi} - S_{\phi\psi}^{(\ell)}$. In fact, the error operators are themselves Cotlar-type operators of the form

$$(4.3) \qquad f \mapsto \int_{\mathbb{R}^n} f(y) \int_H U^*(z) F_\Delta(z) \phi(y) U^*(z) F_\Delta(z) \psi(x) \, dz dy$$

where

$$(4.4) \qquad F_\Delta(z) = \sum_{j,k} \chi_\Delta(z\lambda_{j,k}^{-1})(I - U(z\lambda_{j,k}^{-1}))$$

with $\Delta = \Delta^{(\ell)}$. As mappings on spaces of molecules the norms of these operators just depend on the uniform estimates for $\|(I - U(z.\lambda_{j,k}^{-1}))\psi\|_{\mathcal{M}_\gamma}$, $z \in \Delta_{j,k}$, and corresponding estimates with ϕ. These estimates are given by Corollary 4.1. The boundedness of S_* then hinges largely on some basic general facts about Calderón-Zygmund operators. It is best to state these results in a general form.

We begin with a family of singular integral operators $T, T_1, T_2, \ldots$. All operators T in this section will be assumed to belong to CZO(δ) and to be continuous on $L^2(\mathbb{R}^n)$. The *Calderón-Zygmund norm*, $\|T\|_{CZ}$, associated to T is the larger of the $\mathcal{L}(L^2)$ norm of T and the optimal constants appearing in the kernel estimates for K.

We wish to consider a sequence T_j of such operators in terms of approximations of T.

(4.5) DEFINITION.
(i) A family $T_1, T_2, \ldots$ as above is a uniform family of Calderón-Zygmund operators if $C_{\{T_\ell\}} = \sup\{\|T_\ell\|_{CZ}\} < \infty$.
(ii) $\{T_\ell\}$ approximates T in the strong Calderón-Zygmund sense if the error operators

$$\mathcal{E}_\ell = T - T_\ell$$

satisfy

$$E_{\{\mathcal{T}_\ell\}} \;=\; \sum_{\ell=1}^{\infty} \|\mathcal{E}_\ell\|_{CZ} < \infty.$$

Given a uniform family of Calderón-Zygmund operators $\{\mathcal{T}, \mathcal{T}_\ell\}$ we associate two maximal operators, defined pointwise by

$$M^{\mathcal{T}} f(x) \;=\; \sup_\ell \mathcal{T}_\ell f(x);$$

(4.6) $$M^{\mathcal{T}}_* f(x) \;=\; \sup_\ell \mathcal{T}_{\ell,*} f(x) \,.$$

Here $\mathcal{T}_{\ell,*}$ is the maximal operator arising from truncation of the kernel K_ℓ of $\mathcal{T}_\ell$. Though this convention violates our use of $\mathcal{T}_*$ for truncation in scale in the Cotlar-type case, this ambiguity is inconsequential below, in view of our equiconvergence result (2.7). The notation $M^{\mathcal{T}}$ is meant to distinguish this operator from the maximal operators $\mathcal{T}_*$ via truncation. The basic lemma, which partially motivates the use of the term *strong approximation*, is then as follows:

(4.7) LEMMA. *Given a uniform family of Calderón-Zygmund operators $\{\mathcal{T}_\ell\}$ which approximate $\mathcal{T}$ in the strong sense. The operator $M^{\mathcal{T}}_*$ is weak type (1,1), that is, for all $f \in L^1$ and $\alpha > 0$,*

$$|\{x : M^{\mathcal{T}}_* f(x) > \alpha\}| \;\le\; \frac{A}{\alpha} \|f\|_{L^1} \,.$$

The constant A is at most a dimensional constant multiple of $C_{\{\mathcal{T}_\ell\}} + E_{\{\mathcal{T}_\ell\}}$.

PROOF. To prove the lemma, fix $\alpha > 0$ and $f \in L^1(\mathbb{R}^n)$. Set $E_\alpha = |\{x : M^{\mathcal{T}}_* f(x) > \alpha\}|$. Notice that E_α is measurable since the sets $\{\mathcal{T}_{\ell,*} f(x) > \alpha\}$ are measurable. We divide $E_\alpha = F_\alpha \cup G_\alpha$ where

(4.8) $$F_\alpha \;=\; E_\alpha \cap \{x : \mathcal{T}_* f(x) > \alpha/2\}; \quad G_\alpha \;=\; E_\alpha \setminus F_\alpha \,.$$

Note that

(4.9) $$|F_\alpha| \;\le\; 2C_n \frac{C_{\mathcal{T}}}{\alpha} \|f\|_{L^1(\mathbb{R}^n)}$$

since $\mathcal{T}_*$ is weak-type (1,1), see e.g., [**Ste2**], pg. 11. On the other hand, since $G_\alpha \subseteq E_\alpha$ it follows that whenever $x \in G_\alpha$, there is some ℓ such that $\mathcal{T}_{\ell,*} f(x) > \alpha$. But since $\mathcal{E}_{\ell,*} > \mathcal{T}_{\ell,*} - \mathcal{T}_*$, and $x \notin F_\alpha$, it follows that $\mathcal{E}_{\ell,*} f(x) > \frac{\alpha}{2}$. Therefore

(4.10) $$|G_\alpha| \;\le\; \sum_{\ell=1}^{\infty} |\{x : \mathcal{E}_{\ell,*} f(x) > \alpha/2\}| \;\le\; 2C_n E_{\{\mathcal{T}_\ell\}} \|f\|_{L^1(\mathbb{R}^n)} \,.$$

The lemma now follows from (4.8)–(4.10). $\qquad\square$

Once the weak-type (1,1) result is proved for the maximal operator $\mathcal{T}_*$ or $\mathcal{K}_*$, L^p continuity can be deduced from Cotlar's inequality, e.g., [**Mey**], p. 250. This avenue is closed to us for $M^{\mathcal{T}}$ because we do not have simultaneous pointwise control of all of the operators $\mathcal{T}_{\ell,*}$. Instead we use a good-λ inequality relating $M^{\mathcal{T}}_*$ to the Hardy-Littlewood operator, from which all of the L^p estimates will follow. The result follows the argument in [**Ste2**], p. 206 ff., almost verbatim once one has Lemma (4.7), so we do not include the proof here.

(4.11) THEOREM. *Suppose that $\{\mathcal{T}_\ell\}$ is a uniformly bounded family of CZO's which approximate the CZO $\mathcal{T}$ in the strong CZ sense. Then there is an $A > 0$ depending on $C_{\mathcal{T}}$ and $E_{\{\mathcal{T}_\ell\}}$ such that for all $c > 0$ and $0 < b < 1$ and for all $f \in L^1$ and $\alpha > 0$:*

$$|\{x : M_*^{\mathcal{T}} f(x) > \alpha;\, Mf(x) \le c\alpha\}| \;\le\; \frac{Ac}{1-b}|\{x : M_*^{\mathcal{T}} f(x) > b\alpha\}| \;.$$

(4.12) COROLLARY. *Under the conditions of (4.11), for each $1 < p < \infty$ and all $f \in L^p$,*

$$\|M_*^{\mathcal{T}} f\|_{L^p} \;\le\; C_p\|f\|_{L^p} \;,$$

and

$$\|M^{\mathcal{T}} f\|_{L^p} \;\le\; 2C_p\|f\|_{L^p} \;.$$

The L^p estimate for $M_*^{\mathcal{T}}$ follows directly from the good-λ inequality by a standard argument. The result for $M^{\mathcal{T}}$ follows from the (uniform) pointwise estimate

$$|\mathcal{T}_\ell f(x)| \;\le\; \mathrm{const.}[\mathcal{T}_{\ell,*} f(x) + |f(x)|] \;,$$

e.g., [**Ste**], p. 36, where, since $\{\mathcal{T}_\ell\}$ is uniform, the constant does not depend on j.

PROOF OF (4.2). Under the hypotheses of the theorem, it immediately follows from the definition of $\mathcal{E}_{\phi\psi}^{(\ell)}$ and from the results of chapter 2 that the sequence $\{\mathcal{S}_{\phi\psi}^{(\ell)}\}$, as operators in $\mathrm{CZO}(\gamma)$, approximate $\mathcal{T}_{\phi\psi}$ in the strong CZ-sense. Therefore $\mathcal{S}_*$, which takes the form of the operator $M^{\mathcal{T}}$ above, is bounded on each L^p by corollary (4.12). This proves the theorem. $\qquad\square$

Appendix A. Proof of Basic Estimates

Here we prove the Basic Estimates used earlier. The first two of these depend only on decay.

PROOF OF BASIC ESTIMATE A(i). By dilation it is enough to show that

$$(A.1) \qquad \int_{\mathbb{R}^n} \frac{1}{(1+|x-v|)^{n+\delta}} \frac{1}{(1+|v|)^{n+\delta}} \, dv \;\leq\; \text{const.} \; \frac{1}{(1+|x|)^{n+\delta}}$$

holds uniformly in x. Set

$$\int_{\mathbb{R}^n} \frac{1}{(1+|x-v|)^{n+\delta}} \frac{1}{(1+|v|)^{n+\delta}} \, dv$$

$$= \left(\int_A + \int_{\mathbb{R}^n \setminus A} \right) \frac{1}{(1+|x-v|)^{n+\delta}} \frac{1}{(1+|v|)^{n+\delta}} \, dv = I_1 + I_2$$

where $A = \left\{ v \in \mathbb{R}^n : |v| \leq \frac{1}{2}|x| \right\}$. Then $|v - x| \geq \frac{1}{2}|x|$ on A, while $|v| \geq \frac{1}{2}|x|$ on $\mathbb{R}^n \setminus A$; consequently,

$$I_1 \leq \text{const.} \; \frac{1}{(1+|x|)^{n+\delta}} \int_{\mathbb{R}^n} \frac{dv}{(1+|v|)^{n+\delta}}$$

while

$$I_2 \leq \text{const.} \; \frac{1}{(1+|x|)^{n+\delta}} \int_{\mathbb{R}^n} \frac{dv}{(1+|x-v|)^{n+\delta}}.$$

Estimate A(i) now follows immediately. $\qquad\square$

PROOF OF BASIC ESTIMATE A(ii). Much as before, set

$$\int_0^\infty \int_{\mathbb{R}^n} \frac{t^\delta}{(t+|x-v|)^{n+\alpha}} \frac{t^\delta}{(t+|v|)^{n+\beta}} \frac{dv\,dt}{t}$$

$$= \left(\int_A + \int_{H \setminus A} \right) \frac{t^\delta}{(t+|x-v|)^{n+\alpha}} \frac{t^\delta}{(t+|v|)^{n+\beta}} \frac{dv\,dt}{t} = I_1 + I_2$$

where $A = \left\{ (v,t) : |v| \leq \frac{1}{2}|x| \right\}$. Then

$$I_1 \leq \text{const.} \int_0^\infty \frac{t^{2\delta}}{(t+|x|)^{n+\alpha}} \left(\int_{\mathbb{R}^n} \frac{1}{(t+|v|)^{n+\beta}} \, dv \right) \frac{dt}{t}$$

$$= \text{const.} \int_0^\infty \frac{t^{2\delta-\beta}}{(t+|x|)^{n+\alpha}} \frac{dt}{t}$$

provided $\beta > 0$. Similarly,

$$I_2 \leq \text{const.} \int_0^\infty \frac{t^{2\delta - \alpha}}{(t + |x|)^{n+\beta}} \frac{dt}{t}$$

provided $\alpha > 0$. Estimate A(ii) now follows easily, the conditions $\max(\alpha, \beta) < 2\delta < \alpha + \beta$ being needed to secure convergence of the last two integrals. $\square$

The remaining estimate all exploit cancellation. Denote by

$$P_{[\delta]} f(x, v) = \sum_{|\alpha| \leq [\delta]} \frac{1}{\alpha!} D^\alpha f(x) v^\alpha$$

the Taylor polynomial of f of degree $[\delta]$ centered at x.

PROOF OF BASIC ESTIMATE B(i). Set

$$\int_{\mathbb{R}^n} \left| f(x + sv) - P_{[\gamma]} f(x, sv) \right| \frac{1}{(1 + |v|)^{n+\delta}} \, dv$$

$$= \left(\int_A + \int_{\mathbb{R}^n \setminus A} \right) \left| f(x + sv) - P_{[\gamma]} f(x, sv) \right| \frac{1}{(1 + |v|)^{n+\delta}} \, dv = I_1 + I_2$$

where $A = \{ v \in \mathbb{R}^n : |v| \leq \frac{1}{2s}(1 + |x|) \}$. By Taylor's theorem,

$$f(x + sv) - P_{[\gamma]} f(x, sv) = \sum_{|\alpha| = [\gamma]} \frac{1}{\alpha!} \left(D^\alpha f(x + \theta sv) - D^\alpha f(x + sv) \right) (sv)^\alpha$$

for some θ, $0 < \theta < 1$, depending an v, s and x. Since

$$1 + |x + \theta sv| \geq 1 + |x| - \theta s |v| \geq \frac{1}{2}(1 + |x|) \quad (v \in A),$$

the Lipschitz condition on the highest order derivatives of $\mathcal{M}_\gamma$-functions thus ensures that

$$|I_1| \leq \text{const.} \|f\|_{\mathcal{M}_\gamma} \left(\int_A \frac{(s|v|)^\gamma}{(1 + |v|)^{n+\delta}} \, dv \right) \frac{1}{(1 + |x|)^{n+2\gamma}}$$

$$\leq \text{const.} \|f\|_{\mathcal{M}_\gamma} \frac{s^\gamma}{(1 + |x|)^{n+2\gamma}},$$

the condition $\delta > \gamma$ being used in estimating the last integral.

To estimate I_2 we deal with the function and Taylor polynomial terms separately. Indeed,

$$\int_{\mathbb{R}^n \setminus A} |f(x + sv)| \frac{1}{(1 + |v|)^{n+\delta}} \, dv \leq \text{const.} \left(\int_{\mathbb{R}^n \setminus A} |f(x + sv)| \, dv \right) \left(\frac{s}{1 + |x|} \right)^{n+\gamma}$$

$$\leq \text{const.} \|f\|_{\mathcal{M}_\gamma} \frac{s^\gamma}{(1 + |x|)^{n+\gamma}} \, .$$

On the other hand,

$$\int_{\mathbb{R}^n \setminus A} |P_{[\gamma]}f(x, sv)| \frac{1}{(1+|v|)^{n+\delta}}\, dv \leq \sum_{|\alpha| \leq [\gamma]} \frac{1}{\alpha!} |D^\alpha f(x)| s^{|\alpha|} \int_{\mathbb{R}^n \setminus A} \frac{|v|^{|\alpha|}}{(1+|v|)^{n+\delta}}\, dv$$

$$\leq \text{const.}\, \|f\|_{\mathcal{M}_\gamma} \sum_{|\alpha| \leq [\gamma]} \left(\int \frac{(s|v|)^{|\alpha|}}{(1+|v|)^{n+\gamma}}\, dv \right) \frac{1}{(1+|x|)^{n+\gamma+|\alpha|}}\ .$$

Now, after a change of variable,

$$\int_{\mathbb{R}^n \setminus A} \frac{(s|v|)^{|\alpha|}}{(1+|v|)^{n+\delta}}\, dv \leq \text{const.} \left(\frac{s}{1+|x|} \right)^\gamma \int_1^\infty \frac{1}{t^{\gamma-|\alpha|}} \frac{dt}{t} \leq \text{const.} \left(\frac{s}{1+|x|} \right)^\gamma.$$

Thus

$$\int_{\mathbb{R}^n \setminus A} |P_{[\gamma]}f(x, sv)| \frac{1}{(1+|v|)^{n+\delta}}\, dv \leq \text{const.}\, \|f\|_{\mathcal{M}_\delta} \frac{s^\delta}{(1+|x|)^{n+2\gamma}}\ .$$

Together these estimates establish B(i), completing the proof. $\qquad\square$

The proof of Basic Estimate B(ii) is exactly the same as that for B(i) with the roles of δ, γ reversed save for the final step in the estimate for I_1. Here the Lipschitz estimate on the highest order derivatives of $\mathcal{M}_\delta$-functions ensures that

$$|I_1| \leq \text{const.}\, \|f\|_{\mathcal{M}_\delta} \left(\int_A \frac{(s|v|)^\delta}{(1+|v|)^{n+\gamma}}\, dv \right) \frac{1}{(1+|x|)^{n+2\delta}}$$

$$\leq \text{const.}\, \|f\|_{\mathcal{M}_\delta} \frac{s^\gamma}{(1+|x|)^{n+2\gamma}}.$$

Bibliography

[Cal]   A.P. Calderón, *Algebras of singular integral operators*, Proc. Sympos. Pure Math., AMS, Providence, RI, 1967, pp. 18-55.

[ChJ]   M. Christ, J.L. Journé, *Polynomial growth estimates for multilinear singular integral operators*, Acta. Math. **159** (1987), 51-80.

[CoM]   R.R. Coifman and Y. Meyer, *Au delà des opérateurs pseudo-differentiels*, Astérisque **57** (1978), 1-185.

[Cot]   M. Cotlar, *A unified theory of the Hilbert transform and ergodic theorems*, Rev. Mat. Cuyana **1** (1955), 105-167.

[CS]   C.K. Chui and X. Shi, *Inequalities of Littlewood-Paley Type for Frames and Wavelets*, SIAM J. Math. Anal. **24** (1993), 263-277.

[Dau]   I. Daubechies, *Ten Lectures on Wavelets*, CBMS-NSF Reg. Conf. Ser. in Appl. Math., No. 61, SIAM, Philadelphia, PA, 1992.

[DJ]   G. David and J.L. Journé, *A boundedness criterion for generalized Calderón-Zygmund operators*, Annals of Mathematics **120** (1984), 371-397.

[DJS]   G. David, J.L. Journé, and S. Semmes, *Opérateurs de Calderón-Zygmund, fonctions para-accrétives et interpolation*, Rev. Mat. Ibero. **1** (1985), 1-56.

[FG1]   Feichtinger, H. G. and Groechenig, K., *Banach spaces related to integrable group representations and their atomic decompositions I*, J. Funct. Anal. **86** (1989), 307-340.

[FG2]   Feichtinger, H. G. and Groechenig, K., *Banach spaces related to integrable group representations and their atomic decompositions II*, Mh. Math. **108** (1989), 129-148.

[FJ]   M. Frazier and B. Jawerth, *A Discrete Transform and Decompositions of Function Spaces*, The Journal of Functional Analysis **93** (1990), 34-170.

[FJW]   M. Frazier, B. Jawerth, and G. Weiss, *Littlewood-Paley Theory and the Study of Function Spaces*, CBMS Reg. Conf. Ser. in Math., No. 79, Amer. Math. Soc., Providence, RI, 1991.

[GHL]   J.E. Gilbert, J.A. Hogan and J.D. Lakey, *Characterization of Hardy Spaces by Singular Integrals and 'Divergence-Free' Wavelets*, preprint.

[Gro]   Grochenig, K., *Describing functions: atomic decompositions versus frames*, Mh. Math. **112** (1991), 1-41.

[Han]   Y.S. Han, *Calderón-type Reproducing Formula and the Tb Theorem*, Rev. Mat. Ibero. **10** (1994), 51-91.

[HaW]   Y.S. Han and G. Weiss, *Function Spaces on Spaces of Homogeneous Type*, Essays on Fourier analysis in honor of Elias M. Stein, Princeton University Press, Princeton, NJ, 1995, pp. 211-224.

[HeW]   E. Hernández and G. Weiss, *A First Course on Wavelets*, CRC Press, 1996.

[Kat]   Y. Katznelson, *An Introduction to Harmonic Analysis*, Wiley and Sons, New York, 1968.

[KKR]   S.E. Kelly, M.A. Kon, and L.A. Raphael, *Local Convergence for Wavelet Expansions*, Journal of Functional Analysis **126** (1994), 102-138.

[Mey] Y. Meyer, *Ondelettes et Opérateurs, vol. I,II*, Hermann, Paris, 1990.

[Mey2] Y. Meyer, *Les Nouveaux Opérateurs de Calderón-Zygmund*, Asterisque **131** (1985), 237-254.

[Sae] S. Saeki, *On the Reproducing Formula of Calderón*, Journal of Fourier Analysis and Applications **2** (1995), 15-28.

[Ste] E. Stein, *Singular Integrals and Differentiability Properties of Functions*, Princeton University Press, Princeton, NJ, 1970.

[Ste2] E. Stein, *Harmonic Analysis*, Princeton University Press, Princeton, NJ, 1993.

[Tch] Ph. Tchamitchian, *Biorthogonalité et Théorie des Opérateurs*, Rev. Mat. Ibero. **3** (1987), 163-189.

[Tor] R.H. Torres, *Boundedness results for operators with singular kernels on distribution spaces*, Memoirs of the AMS, Number 442, (1991), 1-172.

[Tri] H. Triebel, *Theory of Function Spaces*, Monographs in Mathematics, Vol. 78, Birkhäuser Verlag, 1983.

[Uch] A. Uchiyama, *A constructive proof of the Fefferman-Stein decomposition of BMO*, Acta. Math. **148** (1982), 215-241.

Editorial Information

To be published in the *Memoirs*, a paper must be correct, new, nontrivial, and significant. Further, it must be well written and of interest to a substantial number of mathematicians. Piecemeal results, such as an inconclusive step toward an unproved major theorem or a minor variation on a known result, are in general not acceptable for publication. Papers appearing in *Memoirs* are generally longer than those appearing in *Transactions*, which shares the same editorial committee.

As of November 30, 2001, the backlog for this journal was approximately 6 volumes. This estimate is the result of dividing the number of manuscripts for this journal in the Providence office that have not yet gone to the printer on the above date by the average number of monographs per volume over the previous twelve months, reduced by the number of volumes published in four months (the time necessary for preparing a volume for the printer). (There are 6 volumes per year, each containing at least 4 numbers.)

A Consent to Publish and Copyright Agreement is required before a paper will be published in the *Memoirs*. After a paper is accepted for publication, the Providence office will send a Consent to Publish and Copyright Agreement to all authors of the paper. By submitting a paper to the *Memoirs*, authors certify that the results have not been submitted to nor are they under consideration for publication by another journal, conference proceedings, or similar publication.

Information for Authors

Memoirs are printed from camera copy fully prepared by the author. This means that the finished book will look exactly like the copy submitted.

The paper must contain a *descriptive title* and an *abstract* that summarizes the article in language suitable for workers in the general field (algebra, analysis, etc.). The *descriptive title* should be short, but informative; useless or vague phrases such as "some remarks about" or "concerning" should be avoided. The *abstract* should be at least one complete sentence, and at most 300 words. Included with the footnotes to the paper should be the 2000 *Mathematics Subject Classification* representing the primary and secondary subjects of the article. The classifications are accessible from `www.ams.org/msc/`. The list of classifications is also available in print starting with the 1999 annual index of *Mathematical Reviews*. The Mathematics Subject Classification footnote may be followed by a list of *key words and phrases* describing the subject matter of the article and taken from it. Journal abbreviations used in bibliographies are listed in the latest *Mathematical Reviews* annual index. The series abbreviations are also accessible from `www.ams.org/publications/`. To help in preparing and verifying references, the AMS offers MR Lookup, a Reference Tool for Linking, at `www.ams.org/mrlookup/`. When the manuscript is submitted, authors should supply the editor with electronic addresses if available. These will be printed after the postal address at the end of the article.

Electronically prepared manuscripts. The AMS encourages electronically prepared manuscripts, with a strong preference for $\mathcal{A}_{\mathcal{M}}\mathcal{S}$-LaTeX. To this end, the Society has prepared $\mathcal{A}_{\mathcal{M}}\mathcal{S}$-LaTeX author packages for each AMS publication. Author packages include instructions for preparing electronic manuscripts, the *AMS Author Handbook*, samples, and a style file that generates the particular design specifications of that publication series. Though $\mathcal{A}_{\mathcal{M}}\mathcal{S}$-LaTeX is the highly preferred format of TeX, author packages are also available in $\mathcal{A}_{\mathcal{M}}\mathcal{S}$-TeX.

Authors may retrieve an author package from e-MATH starting from
`www.ams.org/tex/` or via FTP to `ftp.ams.org` (login as `anonymous`, enter
username as password, and type `cd pub/author-info`). The *AMS Author Handbook* and the *Instruction Manual* are available in PDF format following the author
packages link from `www.ams.org/tex/`. The author package can be obtained free
of charge by sending email to `pub@ams.org` (Internet) or from the Publication Division, American Mathematical Society, P.O. Box 6248, Providence, RI 02940-6248.
When requesting an author package, please specify $\mathcal{AMS}$-LaTeX or $\mathcal{AMS}$-TeX, Macintosh or IBM (3.5) format, and the publication in which your paper will appear.
Please be sure to include your complete mailing address.

Sending electronic files. After acceptance, the source file(s) should be sent to
the Providence office (this includes any TeX source file, any graphics files, and the
DVI or PostScript file).

Before sending the source file, be sure you have proofread your paper carefully.
The files you send must be the EXACT files used to generate the proof copy that was
accepted for publication. For all publications, authors are required to send a printed
copy of their paper, which exactly matches the copy approved for publication, along
with any graphics that will appear in the paper.

TeX files may be submitted by email, FTP, or on diskette. The DVI file(s) and
PostScript files should be submitted only by FTP or on diskette unless they are
encoded properly to submit through email. (DVI files are binary and PostScript
files tend to be very large.)

Electronically prepared manuscripts can be sent via email to
`pub-submit@ams.org` (Internet). The subject line of the message should include
the publication code to identify it as a Memoir. TeX source files, DVI files, and
PostScript files can be transferred over the Internet by FTP to the Internet node
`e-math.ams.org` (130.44.1.100).

Electronic graphics. Comprehensive instructions on preparing graphics are available at `www.ams.org/jourhtml/graphics.html`. A few of the major requirements are given here.

Submit files for graphics as EPS (Encapsulated PostScript) files. This includes
graphics originated via a graphics application as well as scanned photographs or
other computer-generated images. If this is not possible, TIFF files are acceptable
as long as they can be opened in Adobe Photoshop or Illustrator. No matter what
method was used to produce the graphic, it is necessary to provide a paper copy to
the AMS.

Authors using graphics packages for the creation of electronic art should also
avoid the use of any lines thinner than 0.5 points in width. Many graphics packages
allow the user to specify a "hairline" for a very thin line. Hairlines often look
acceptable when proofed on a typical laser printer. However, when produced on a
high-resolution laser imagesetter, hairlines become nearly invisible and will be lost
entirely in the final printing process.

Screens should be set to values between 15% and 85%. Screens which fall outside
of this range are too light or too dark to print correctly. Variations of screens within
a graphic should be no less than 10%.

Inquiries. Any inquiries concerning a paper that has been accepted for publication should be sent directly to the Electronic Prepress Department, American
Mathematical Society, P. O. Box 6248, Providence, RI 02940-6248.

Selected Titles in This Series

For a complete list of titles in this series, visit the
AMS Bookstore at **www.ams.org/bookstore/**.